Anth | Tribal

500gm

AF529218

DEVELOPMENTAL ANTHROPOLOGY

By the Same Author

Manavshastriya Anusandhan Vidhi Awam Taknik
(Anthropological Research Method and Technique)

Jatil Samaj Ka Manavshastra
(Anthropology of Complex Society)

Bhartiya Manavshastra
(Indian Anthropology)

Bhartiya Janajatiya Sanskriti
(Indian Tribal Culture)

Manav Shastriya Sidhanta: Shastriya and Adhunik
(Anthropological Theory: Classical and Modern).

About the Author

Gaya Pandey (*b.* 1954) is a senior faculty in Anthropology at Ranchi University, Ranchi (Jharkhand). Ever since his joining as faculty member in this institution (1985), he has been actively engaged in fieldwork, teaching and research. He has to his credit dozens of articles published in journals of national and international repute. He has co-authored nine books on different areas of Anthropology. Besides, he has published many course books in Anthropology in Hindi medium including *Manavshastriya Anusandhan Vidhi Awam Taknik* (Anthropological Research Method and Technique), *Jatil Samaj Ka Manavshastra* (Anthropology of Complex Society), *Bhartiya Manavshastra* (Indian Anthropology), *Bhartiya Janajatiya Sanskriti* (Indian Tribal Culture), and *Manav Shastriya Sidhanta: Shastriya and Adhunik* (Anthropological Theory: Classical and Modern).

His books published under co-authorship include *Sun Workship in India* (1992), *History of Anthropological Thought* (1993), *Manavshastriya Vicharak Ewam Vichardharayen* (2000), *Forest and Tribals* (1997), *Artisan Tribes of Bihar and Bengal* (2001), *Janajatiya Vikas* (2001), *Vikasatmak Manav Vigyan* (2001), *Tribal Development in India* (2002), and *Samajik Sanskritik Manav Shastra* (2005).

Developmental Anthropology

Gaya Pandey

CONCEPT PUBLISHING COMPANY, NEW DELHI-110059

Cataloging in Publication Data—DK
Courtesy: D.K. Agencies (P) Ltd. <docinfo@dkagencies.com>

Pandey, Gaya, 1954-
Developmental anthropology / Gaya Pandey.
p. cm.
In Indian context.
Includes bibliographical references (p.)
Includes index.
ISBN 13: 9788180695704
ISBN 10: 8180695700

1. Applied anthropology—India. 2. Political anthropology—India. 3. Social planning—India. 4. India—Economic policy. 5. Rural development—Government policy—India. I. Title.

DDC 306.095 4 22

All rights reserved. No part of this work may be reproduced, stored in a retrieval system, or transmitted in any form or by any means electronic, mechanical, photocopying, micro-filming recording or otherwise, without the prior written permission of the copyright owner and the publisher.

ISBN-13: 978-81-8069-570-4 (HB)
ISBN-10: 81-8069-570-0 (HB)
ISBN-13: 978-81-8069-571-1 (PB)
ISBN-10: 81-8069-571-9 (PB)

First Published 2008

© Gaya Pandey

Published and Printed by
Ashok Kumar Mittal
Concept Publishing Company
A/15-16, Commercial Block, Mohan Garden
New Delhi-110059 (INDIA)

Phones : 25351460, 25351794
Fax : 091-11-25357103
Email : publishing@conceptpub.com

Dedicated to
most reverent father Late Ram Pravesh Pandey and mother Smt. Ansuiya Devi whose sacred feet and blessings have inculcated the Sanskar of loyalty towards work and profession in me

Preface

Man is only being on this earth who has thought, planned and made development ever since he emerged on this planet from his animal ancestry. In course of history, he passed through stages of savagery, barbarism and civilisation. Series of inventions were made and he became civilised citizen of a nation. After emergence of State, the responsibility of the development in different areas is owned by the Government of the concerned state. Development Anthropology is one of the branches of anthropoly which aims at studying and researching on meaning, concept, theories, models, approaches, policies, strategies and programmes of development. The present book is an scholarly attempt to present these aspects of development before readers, scholars and students.

Undoubtedly, there are books dealing with different aspects of development, but readers do not find all those aspects together in a single book form. As a result, they are facing trouble in their study. Attempt has been made to solve these problems of scholars and students in this book.

The book could be completed with the help extended by a number of persons. Hence it is my moral responsibility to extend my sincere gratitude to them. I have completed this book after going through a good number of books, articles, reports and other reading materials written by my *Guru* (Teachers), Colleagues, Friends and Scholars. I express my gratitude to all of them collectively and individually.

I am thankful to my mother Smt. Anasuiya Devi, wife Smt. Indu Devi, daughters—Sudha, Suman, Sarita and Anu and my son Shekhar for their constant support and encouragement.

I am also thankful to Sri Ashok Kumar Mittal and his team of Concept Publishing Company, New Delhi for printing the manuscript of the book with meticulous care.

The real evaluator of a book are students, scholars and teachers. If they are satisfied with presentation under the book, this will be the great success of author and book.

Gaya Pandey
Department of Anthropology
Ranchi University
Ranchi-834008 (Jharkhand)

Contents

1

Developmental Anthropology
An Introduction

Definition

Anthropology is total or wholistic study of man. This science is total study of man because it studies man as a whole. The study of man as a whole includes the biological, cultural, social, economic, political, religious, asthetic, artistic, linguistic, historical, medical and developmental beings. Considering the wholistic nature of anthropology, Developmental Anthropology can be defined as a sub-branch of anthropology which makes detail and comparative study of developmental aspects of man. Man is the only animal on this earth who thinks about development, plans about development, formulates and implements schemes about development, makes evaluation of the development and brings changes in the policies, plans, and strategies of development. Thus, man is the only developmental beings on this earth and developmental anthropology makes study of man as a developmental beings.

Historical Background

In early times when man emerged from his animal ancestry, he was completely dependent upon nature particularly forest for his survival. He was not producer, rather he was consumer of goods provided by the nature. He had to struggle hard to survive in glacial, interglacial, pluvial and interpluvial environments. In order to survive in this changing environments of Pleistocene period man took help of stone. He started making tools from stone pieces by different techniques of flakings. This gradually gave rise to stone tool industries and cultures. In the beginning stonetools were not very beautiful. They were big with a butt end and a working end. They were named as handaxe. This tool industry and culture prevailed in entire lower Palaeolithic period. In the middle Palaeolithic period, scrapper tool industries and cultures came into existence. In upper Palaeolithic period blade and bone tool industries and culture were developed. Art culture and magician-cum-specialist culture also developed in this period. In Mesolithic period, microlithic and bow-arrow industries and culture developed. In Neolithic period, polished and grinding tools, different types of celts, agriculture, domestication of animals, weaving, pottery, houses, villages, and megaliths were developed. In Chalcolithic age, copper tools, plough, yoke, spade, spadle, wheeled cart, potter's wheel, money etc. were invented.

In Bronze age, urban revolution took place. This revolution gave rise to civilisation. State came into existence. Temples were built. Priests and traders came into existence. Script, numerals, accountancy, standard weights and measures, literature, solar calendar, seals, capidary, glazing technology and science were invented.

In ancient period, state, administration, army, education and medicine were used in developed form. In Medieval period, along with these traits, paper industries and dye industries were invented. In modern British period, scientific and industrial revolution took place. As a result of this, modernisation took place. The process of modernisation gave rise to modern aeroplane, ship, railway, motorcar, truck, road transport, television, telephone, radio, education, medicine, law, government, army, war weapons, democracy and capitalism.

In post-modern period, knowledge revolution has taken place. This process has given birth to global economy, market, culture and politics.

From the above account, it is quite clear that man has been able to make development in different areas right from pre-historic time. Through different times and processes he has come into the present state of development. Thus, development of man as a whole is a continuous process.

Origin of Developmental Anthropology

Developmental Anthropology as a branch of Anthropology is recent in origin. Although the development of man as a whole has a long history and story to present before us, but the origin of developmental anthropology is not so old. Its origin can be traced from the rise of independent countries from the British imperial power and emergence of International developmental agencies like UNO, World Bank, IMF, UNICEF, CARE, RED Cross, ILO, WHO etc. In fact, the origin of developmental anthropology can be traced after the World War II. After World War II, developed nations began to think that the problems of development of developing nations are quite different. The undeveloped and developing countries were facing problems like drinking water, house, rural roads, health, education, nutrition, gender discrimination, casteism and communalism. In these countries people belonging to weaker sections particularly women, children, labour, STs and SCs were deprived, exploited and oppressed. There was urgent need to protect them from deprivation, exploitation and oppression.

Our democratic Government is engaged in protecting the weaker sections and bringing them into mainstream of development soon after independence. Right from the very First Five Years Plan, constitutional safeguards have been provided to them. Development policies and programmes have been formulated, implemented, evaluated and changed to bring them in the mainstream of the nation. With these developments, developmental anthropology came into existence.

Subject Matter (Scope) of Developmental Anthropology

The scope of developmental anthropology includes the teaching and research on the following subject matters:

1. Meaning and concept of development.
2. Theories and modes of development.
3. Types of development.
4. Areas of development.
5. Indicators or parameters of development.
6. Impact of development.
7. Development and change.
8. Development planning, policies and strategies.
9. Formulations of developmentplans and projects.
10. Implementation of development plans and projects.
11. Monitoring and evaluation development plans and projects.
12. Constitutional safeguards regarding the development.
13. Social-cultural factors of development.
14. Barriers of development.
15. Agencies related to development.
16. Industrial development, displacement and rehabilitation.
17. Development and social justice.
18. Panchayati Raj and development.
19. Child development.
20. Women development.
21. Scheduled Castes development.
22. Schedule Tribes development.
23. Human Resource development.
24. Health development.
25. Rural development.
26. Knowledge and communication development.
27. Role of anthropology in development.
28. Role of anthropologists in development.

Applied Anthropology, Action Anthropology and Development Anthropology

In anthropological literatures, one finds at some places the use of Applied Anthropology, Action Anthropology and Developmental Anthropology interchangeably. But there is difference between the nature of these three sub-branches of anthropology. Therefore, it will be the fitness of the things to present here the difference between Applied Anthropology, Action Anthropology and Development Anthropology.

Applied Anthropology

The term Applied Anthropology is legacy of British anthropologists. Applied anthropology deals with the application of anthropological knowledge in solving problems of human society such as racism, casteism, communalism, terrorism, war weapons, ethnocentrism, cultural relativism, medicine, law, administration and development. By applying anthropological knowledge, these problems of human society can be solved. Anthropologists have solved a number of ethnic problems on the basis of their anthropological knowledge.

Action Anthropology

The term Action Anthropology is American legacy. It was used by SolTax for the first time. Action anthropology deals with the study of anthropological knowledge in action. Anthropological knowledge to solve the problems of human societies is fruitless without bringing it in action. Anthropologists have attempted to rehabilitate and develop a number of ethnic groups on the basis of their anthropological knowledge in action.

Development Anthropology

The term Development Anthropology is legacy of anthropologists working with developmental processes, theories, models, planning, strategies and performance in developing countries. This science deals with the study of development perspectives in totality after World War II.

Study Designs of Developmental Anthropology

Developmental anthropology makes use of following designs in its study and research:

1. *Baseline Study Design*: In order to know the impact of development baseline study design is formulated and implemented. This design provides us with the baseline data of development.

2. *Impact StudyDesign*: On the basis of impact study design one attempts to know the positive or negative impact of development programmes. This design is implemented after a gap of time of implementing baseline study design. The gap of time may be one, two, three, four, five, years.

3. *Experimental Study Design*: This design is also prepared and implemented in order to know the impact of development. In this type of study design, two groups are chosen for the purpose of study. One group is called control group and the other group is termed as experimental group. The experimental group is chosen on the basis of development experiment going on in the group. The control group remains away from the experimentation of development programme. The difference in the status of these groups reveals impact of development.

4. *Evaluative Study Design*: This type of study design is prepared in order to know the end result after the implementation of a development programme. On the basis of this research design developmental anthropologists attempt to reveal performance and success—failure of schemes of development. It also makes cost-benefit analysis of development programmes meant for the people.

5. *Descriptive Study Design:* This study designs attempts to describe the development programmes implemented for the benefit of the people.

2

Concept of Development

The term development is a value based term. The term development is valuable because it is used to distinguish developed ones from developing and undeveloped ones or to uplift the weaker sections and undeveloped areas. There are several human groups or areas which are undeveloped on the basis of several parameters of development. Therefore, for those human groups which are not developed, this is a need to develop them. The term development is conceived as valuable for policymakers, planners, implementers, evaluators and beneficiaries. Everybody accepts development gladly and takes it in positive sense. The term finds value at individual, family, society and national level.

Impact of development has devided human groups, regions and nations into three categories: (i) Developed one, (ii) Developing one, and (iii) Undeveloped one.

America, England, Japan, etc.: are termed as first world countries because they are developed ones. India comes under developing countries. It is also known as third world. Some African countries come under the category of undeveloped countries. The criteria of keeping one under the category of developed, developing and undeveloped is parameters of development such as income, education, health, housing, sanitation drinking water etc.

In India forward castes are placed in developed categories. Backward castes have been placed under developing categories and STs and SCs have been placed under undeveloped category.

Similarly, in our country metropolis like Kolkata, Mumbai, Chennai and Delhi represent the first world. Other cities represent second world and Towns and Villages represent the third world.

Thus, according to the concept of development, it is a valuable term. It is conceived positively. It is understood on the basis of parameters. It devides countries, regions and people into developed, developing and undeveloped categories. It emphasises upon the need to develop the undeveloped ones, by planning and programming.

Meaning of Development

The term development caries the following meaning:

1. Increase in income
2. Rise in social prestige
3. Aesthetic and artistic achievement
4. Attainment of social virtues
5. Attainment of education and knowledge
6. Increase in tolerance
7. Increase in military power and weapons
8. Achievement of science and technology
9. Good health and nutrition
10. Availability of good sanitation
11. Availability of safe drinking water
12. Good administration and governance
13. Justice to all
14. Attainment of human rights
15. Protection and conservation of environments
16. International brotherhood
17. Achievement of transport, travelling and tourism facilities
18. All be happy
19. End of deprivation, exploitation and oppression
20. Control over anti-social elements and acts
21. Increase in communication facilities
22. Achievement of different resources such as natural water, human, etc.

23. Physical, cultural and economic rehabilitation of displaced and disadvantaged groups
24. Desired change by planning
25. Goal oriented change by planned action
26. Freedom from bondage of all kind.

Definition of Development

The term definition can be defined as a positive change in the present status of an individual, group, region or nation through developmenal intervenations. The positive change is brought not spontaneously but by planning policies, strategies, implementation, monitoring and evaluation of development programmes.

Parameters of Development

For judging the development, certain parameters have been identified. The chief parameters on the basis of which development is judged are as follows:

1. ***Economic Parameters***: Economic prosperity is one of parameters of development. Individual, family or nation whose income is very high comes under the category of developed ones. The countries which have very low percentage of BPL Family are termed as developed one. Economic parameters of a nation includes per capita income, GDP, GNP, NI etc.

2. ***Social Parameters***: Social prosperity is also one of parameters of development. The nations where social prosperity is found is termed as developed one. Social prosperity is counted on the basis of peace, tolerance, co-operation, help, duty, responsibility, brotherhood, enjoyment of status and rights etc.

3. ***Cultural Parameters***: Cultural achievement is also one of parameters of development. Cultural achievements in the form of art, crafts, music, drama, dance, song, hospitality etc. are cultural parameters of development.

4. ***Religious Parameters***: Religion is also one of parameters of development. Religious tolerance, harmony, peace, co-

operation, benevolence etc. are parameters of religious development.

5. ***Political Parameters***: Political institutions are also one of parameters of development. Strong political sovereignty, democracy, military force, modern war equipment and foreign policy are political parameters of development.

6. ***Psychological Parameters***: Psychological traits are also one of parameters of development. Healthy mind, positive attitude, democratic opinions, national feelings, national characters etc. are psychological traits of development.

7. ***Health Parameters***: Health is one of parameters of development. The society or nation which has good nutrition, health, medical facilities, safe drinking water, clean house and environment comes under the category of developed one.

8. ***Education Parameters***: Education makes development of human resources. It makes people aware of human rights, health deficiencies, exploitation, oppression, deprivation, communicable diseases, medicine and various development programmes. It also provides modern job opportunities. Education plays valuable role in development.

9. ***Communication Parameters***: Communication is also one of parameters of development. In absence of proper communication development cannot be thought. Communication is of various types such as road communication, transport communication, railways communication, waterways communication, airways communication, telephone communication, television communication, internet communication etc.

Areas of Development

Development is applied in various areas. The areas where development is applied can be of following types:

1. *Social Area*: The social area of development includes child development, women development, aged development, youth development, OBCs development, SCs development and STs development.

2. ***Economic Area***: The economic area of development includes horticultural, agricultural, irrigational, industrial, cottage industrial, mining, labour and life stock rearing development.

3. ***Cultural Area:*** Cultural area of development reveals development of arts, crafts, song, music, drama, dance, plays, sports, hospitality etc.

4. ***Religious Area***: The religious area of development includes development of religious centres, tolerance, peace, harmony, voluntary organisations.

5. ***Political Area***: Political area of development includes the development of panchayat, thana, police, court, block, district, commissionary, state and nation.

6. ***Geographical Area***: Geographical area of development includes the development of hills, forests, deserts, plains, sea coasts, islands, rural, urban and metropolis.

7. ***Educational Area***: Educational area of development includes the development of infrastructure and human resource, sports, etc.

8. ***Psychological Area***: Psychological area of development includes development of personality, attitude, opinion, equal opportunity etc.

9. ***Health and Hygiene Area***: Health and hygiene area of development includes the development health sub-centres, health centres, block hospital, district hospital, medical colleges and hospitals, different types of medicines, safe drinking water, sanitation and environment cleanliness.

10. ***Communication Area***: The communication area of development includes development of roadways, waterways, air ways, electronic media, telephone, television, internet etc.

11. ***Tourism Area***: The tourism areas of development are development of tourist centres, guest houses, hotels, lodges, *dharamashalas*.

12. ***Fuel Area***: Fuel area of development includes development of wood energy, cowdung energy, biogas energy, coal energy, electric energy, water energy, wind energy, solar energy, petroleum energy, atomic energy etc.

Types of Development

Although development is defined as result oriented change by planned action, but it can be classified into following types:

1. Natural or Spontaneous Development.
2. Planned Development.
3. Pleasant Development.
4. Painful Development.
5. Sustainable Development.
6. Imposed Development.
7. Participatory Development.

1. Natural Development

Some Developments are product of natural events. They develop due to natural events spontaneously. For example, creation of sea, river, well, mountain, continents, islands, desert, savana, forests, cotton soil, plain etc. are not man-made. They are natural creation, not the creation of man.

2. Planned Development

When development is goal oriented through certain planned human actions, it is termed as planned development. For example, rural development, urban development, metropolis development, child development, youth development, women development, education development, health development, communication development etc. These are all examples of planned development.

3. Pleasant (Painless) Development

When development through planned action brings pleasure among the beneficiaries, it is known as pleasant development. Education, employment, health, sanitation, communication development etc. are examples of pleasant development.

4. *Painful Development*

When development through planned human action brings dissatisfaction and displeasure among the people, it is known as painful development. Displacement due to industrial development is an excellent example of painful development.

5. *Sustainable Development*

Development should not be temporary and of short span. It should be of long span and sustainable. Development can be sustainable when it is pro-poor, pro-women, pro-children, pro-youth, pro-aged, pro-ecology, pro-environment, pro-local community and pro-human rights. It is anti-deprivation, anti-exploitation, anti-oppression and anti non-enjoyment of human rights. Thus, sustainable development can be defined as protection of children, women, Aged, SCs, STs, OBCs, labourers etc. from deprivation, exploitation and oppression through planned human actions. It can also be defined as conservation of land, forests, wild life, hills, mountains, rivers, seas, islands, waterfalls, tourist centres and rights through planned human action.

Development is not only the birth right of anyone generation, group, region, ecology and environment. Every generation has equal right of development. Children, youth and aged have right of their development. Men and women both have right of development. G.Cs, OBCs, STs, SCs all have right to develop themselves. Similarly Hindus, Muslims, Sikhs, Jews, Jaina, Buddhists, all have right of development. Similarly, land, forests, wells, plains, deserts, sea coasts, islands, river, mountains, birds, animals, wild life situated in different ecology and environment have right of conservation, protection and development.

Sustainable development believes in balanced and overall improvement in society. The concept of sustainable development emerged due to dissatisfaction and unrest shown by the people against on going development projects. Those projects were imposed upon them against their will. As a result people began to face problems related to employment, displacement and rehabilitation. These problems compelled the people to organise themselves and launch movement against deforestation, construction of dam and establishment of hydel project. Such development activities had raised the problems of unemployment, exploitation, oppression, and deprivation among them. Social scientists, environmentalists, human right protectionists, wild life conservationists etc. have supported the movement of the people against those developmental projects.

It has been observed that man has started consuming the nature particularly land, forest, mountain, rivers etc., ruthlessly by modern technology. The ruthless exploitation of natural objects by scientific technology brought the problems of unemployment, exploitation, oppression, deprivation, natural calamities like flood, earthquakes, epidemic, displacement, rehabilitation and extinction before the local people. The forest cover decreased due illegal felling of trees and cutting of trees by forest departments for furniture, building construction, paper industries and plywood industries. Decrease in forest cover brought the problem of unemployment before the local people who subsisted on food collection and gathering economy. Due to unemployment, exploitation of children, women, aged and labourers increased. Children were employed as child labour on low wages. Women were employed on low wages and exploited physically. Labourers were made bonded labourers and exploited in wage payment. Aged people began to face the problem of social security. Local people were deprived of their human rights.

Similarly, construction of dam, establishment of hydel project and installation of industries brought the problem of submergence, displacement, rehabilitation and extinction before the local people. Such developmental activities also brought the problem of flood, earthquakes and calamities. Coal washeries and industrial waste brought the problem of water and land conservation. They also brought the problem of a number of diseases. Ruthless exploitation of coal, diesel and petrol brought the problem of air pollution. Cities and metropolis faced the problem of noise pollution. There is no doubt that per capita income of certain families involved in Government Jobs were increased, but there was not balanced and over all improvement in quality of life of local people. Rather they began to face the problem of extinction. Thus development brought devastating effect upon the life and culture of the local people.

Considering the devastating impact of development project the idea of sustainable development emerged. Under the concept of sustainable development, emphasis was placed on balanced development and over all improvement in quality of life. According to this concept of development, the planners of development model must plan such a development model which could protect the environment and all people and area could achieve its benefit. Sustainable development favours both macro and micro level development. It lays emphasis on natural balance, protection of environment, growth of economic and welfare of all classes and areas.

For the success of sustainable development participatory and decentralised resource management is needed. In this type of development local people are made participatory group. On the basis of their participation, development programme is formulated and implemented. The responsibility of resource management is also given to the participatory group. This development lays emphasis much upon the rehabilitation of the displaced group. Because only rehabilitation through service, compensation and resettlement cannot solve the problem of displaced group. The displaced groups need integrated development so that they could survive in new environment.

For sustainable development and sustainable rehabilitation of displaced groups, the following points should be taken into consideration:

1. The exploitation of natural resources should not cause ecological imbalance.
2. Emphasis should be placed on renewable of natural resources for future generations.
3. Adequate amount of income from the development project should be spent on the welfare of local people.
4. Uprooted families must be physically rehabilitated with basic amenities of life.

5. Besides physical rehabilitation cultural rehabilitation of displaced group should also be taken in to consideration,
6. Follow up action should also be taken in to view after rehabilitation.
7. Sustainable income generating schemes should be implemented.

6. Imposed Development

When development project or scheme is imposed upon the local people by the Government, it is termed as imposed development. In such type of development planning local people have no participation. Local people feel such type of development imposed upon them. That is why such type of development has to face the resistance of local people. Local people oppose these schemes because they have been imposed upon them without taking their needs in to consideration. For example, giving milch cattle to wandering tribes, installation of industries and hydel project in tribal areas etc. This type of development is result of top down policies of the Government.

7. Participatory Development

When development projects or schemes are formulated and implemented by the participation of local people, it is termed as participatory development. This type of development is result of down top policies of the Government. A number of rural, tribal and women development programmes are being formulated, implemented, monitored and evaluated by the participation of the people.

Progress and Development

It will be the fitness of the thing to present here the difference between the concept of progress and development.

Progress is a continuous process. It reveals stage after stage development in unilineal sequence. That is why progress-oriented development is termed as evolution. Progress reveals that higher forms develop in place of lower forms.

After development of higher form lower form remains as survival and it represent the early condition. Progress creates degeneration. In course of progress, degeneration of traits takes place, but progress is not concerned with degeneration of traits.

Development on the other hand, like progress does not take place spontaneously. Development is planned and goal oriented. It is induced by planning, implementation of schemes, introduction of inputs, monitoring, evaluation and follow up action. It is time bound. It may or may not be followed after lapse of plan period. In this way, we find that progress is spontaneous and development is induced and planned by human activities for welfare of disadvantaged group.

Development and Growth

It is also necessary here to present the difference between growth and development because in literature we find the use of the term growth and development inter-changeably. But there is profound difference between the concept of growth and development.

Growth is a physical quality. It can be measured and weighted in terms of quantity. For example, the growth of a child is measured in terms of length and weight. The growth of population is measured in terms of percentage increase in a particular period. Economic growth is a measured in terms of per capita income, GDP, GNP, NI, etc. Educational growth is measured in terms of literacy rate. Thus, growth is concerned with improvement in one part or sector.

Development, on the other hand, is conceived as balanced and over all improvement in quality of life. The growth rate or per capita income a group can be increased by effective measures without considering the other aspects of life and culture. It may not be called development. Development includes over all and balanced improvement in quality of life.

Development and Change

Development and change are closely related. Development is introduced in order to get desired goal. Thus, desired goal oriented change is called development. All changes are not development. The changes brought by the introduction of schemes, plans, project to achieve desired goal are called development. In order to get desired goal, developmental inputs are induced. The induction of development inputs results in the achievement of desired goal or change. In case of development, the change is not spontaneous. Rather here change is induced by planning and programming, monitoring, and evaluating and following up schemes. Thus, in case of development, first comes developmental inputs and then comes goal achievement. Goal achievement shows change in the past status of the group. The changed status reveals development. In this way, development and change are closely related. Both are product of a series of processes like planning, formulation of strategies and schemes, implementation of schemes, monitoring of scheme, evaluation of scheme, follow up of scheme and attainment of goal, change in status of concerned group and development as end product.

Development and its Impact

As referred to earlier, development is the end product and it has to pass through a series of processes. These processes involve steps like identifying the area, or beneficiaries, formulating policies and programmes, implementing programmes, monitoring, evaluation, and follow up action. By following these steps or processes, the end product desired goal or development is achieved. The development is value oriented concept. The development is evaluated on the basis of cost—benefit analysis and success or failure of the scheme. When development inputs bring success in achievement of goal, the impact of development is taken as positive. But when the development inputs become unsuccessful in achieving the goal, the impact of development is taken as negative. In this way, although development is a value-based concept, but its impact may be positive or negative.

Dilemma of Development

As referred to earlier that development is a value-based concept. It is conceived as goal-oriented change, attainment of desired goal, positive impact, welfare of the people, balanced improvement and achievement of quality of life. But empirical experiences on development reveals the fact that development also brings negative impact. It creates imbalance in nature. It uproots a number of families and forces them for extinction. It becomes failure. It has enhanced the corruption and it has brought adverse effect upon the quality of life. Thus, Paradoxes of quality in theory and practice of development reveal dilemma of development. Some dilemma about development are as follows:

1. Development is value-based concept but it has negative impact, too.
2. Development is goal-oriented change but it becomes failure, too.
3. Development is conceived as welfare of people, but it has devastating effect, too.
4. Development is welcomed by all, but it is criticized for increase of corruption.
5. Development is taken as balanced improvement, but it has created imbalance in nature and man.
6. Development brings improvement in quality of life, but it has affected the quality of life adversely, too.
7. Development is planned and implemented for the benefit of people, but it is opposed by the people, too.
8. Development is for beneficiaries, but it has been proved more beneficial for implementing anthorities.
9. Development has divided people as FCs, OBCs, SCs, STs, but in each category all are not developed equally.
10. Development is taken as freedom from bondage of all kinds but it has increased bondage of various kinds.
11. Development is enjoyment of human rights, but it also leads to non-enjoyment of human rights.

Development is Value-based Concept, but it has Negative Impact, too

Development is considered valuable because it is expected to make people developed. It is also expected to bring improvement in quality of life. But its dilema is that it also brings negative impact. Educational development has created the problem of unemployment. Educated unemployed youth do not want to take up their traditional occupation. They are becoming victim of intoxication, drug abuse, frustrations, murder, suicide etc.

Economic development has created a wide gap between the rich and the poor. The poor are becoming more and more poor and the rich are becoming more and more rich. It has caused rural to urban migration. It has created the problem of slum dwelling, unauthorized occupation of land, theft, dacoity rape, murder, suicide, infantycide, foeticide, and old age security of aged. It has also created the problem of exploitation, oppression and deprivation.

Industrial development has created the problem of water pollution, air pollution, soil pollution, sound pollution, displacement, uprooting of families, rehabilitation, extinction etc.

Material development has caused the problem of gender discrimination, foeticide, infanticide, bride burning, dowry death etc.

In this way, there are several examples which show the negative impact of development.

Development is Goal-oriented, but it becomes Failure, too

It is generally conceived that development is goal-oriented. In order to achieve goal, target group is subjected to developmental inputs. The development inputs bring success only when they are implemented, monitored and followed up properly. Participation of target group is also a must. Without the participation of target group, goal cannot be achieved and success cannot be obtained.

Evaluation reports dealing with the evaluation of several development schemes reveal that programmes of development could not achieve success due to not proper implementation, monitoring, follow up action and participation of target group or local people.

The Birhor of Jharkhand abandoned their rehabilitation colony because it was not serving their cultural and psychological needs.

Community laterine scheme of women could not achieve success because it was not providing them occasion to discuss domestic problems among them as they used to do at the time of defication in the field in early morning and late evening.

Rehabilitation scheme of bonded labour could not achieve success due to lack of monitoring and follow up action.

Rehabilitation scheme of displaced families of Bokaro and Hatia industrial complex could not achieve success and a good number of them have become extinct.

Thus, we see that development is goal oriented and brings desired change, but it becomes failure, too.

Development is Conceived as Welfare of the People, but it has Brought Devastating Effect too

There is no doubt that development is associated closely with the welfare of the people. It has been beneficial for the welfare of disadvantaged group like STs, SCs,OBCs, children and women. But its dilemma is that it has brought devastating effect on the life of the people, too.

For the purpose of development of the people and area, industrial complexes were established by our Government at HEC Hatia, Ranchi, Bokaro steel plant, Jamshedpur steel plant, Patratu and Chandrapur thermal station etc. in Jharkhand. Similarly, industrial complexes were established at Raurkela in Orissa, Bhilai in Chhattisgarh and Durgapur in West Bengal. For installing industries lands were obtained from the people. They were displaced and uprooted from their original place of residence. For them rehabilitation scheme was launched. But evaluation study of rehabilitation schemes for displaced and uprooted families of these industrial complexes reveals that uprooted families were not rehabilitated properly. Some of them have become slum dwellers, while some others have become extinct. Serving families have been forced to live like beggars. From land owners, they become landless labourers.

Some are serving on rickshaw pulling and wage earning. Some are involved in criminal activities, while some families vanished due to death.

River valley projects have also brought devastating effect on life of the people. They have brought the problem of uprooting of villages and plant and animal species. They have caused natural calamity like flood, earthquakes, epidemics, starvation, etc.

Account presented above presents the dilemma of development that it is undoubtedly related to welfare of the people, but it has brought devastating effect on the life of people, too.

Development is Welcomed by all, but it is Criticised for Increasing Corruption

Everybody likes development. Everybody wants to make his development and likes to live in developed society. That is why everybody welcomes the proposal of development. But unfortunate scene related to development is that it has increased the corruption. That is why it is criticised by the people.

Bribery and loot of money in the name of development is the most unfortunate and worst practice prevalent in our society. The work of development is done by the agencies. The agencies may be Governmental or non-Governmental. Again non-Governmental agencies may be local, national and international. The development is not independent. It is dependent upon these agencies and their dedication commitment and honesty.

The evaluation reports dealing with several development schemes like digging wells, installing handpumps, digging ponds, making banks, road construction, bridge construction, distribution of milch cattle, goats, pigs, medicines at health centre etc. have revealed the prevalence of large scale corruption in the form of bribery and loot of public money. Ministers, bureaucrats, contractors, clerks, implementing officials and NGOs are involved in bribery and corruption.

Development fund is distributed among them. The development work is shown just on paper. In some cases 100 per cent funds are grabed by the ministry, development contractors, officials, bureaucrats and NGOs. Out of a rupee, only 10 paisa are spent, rest are grabed.

NGOs have proved boon for the concerned ministers. NGOs are created and gifted to wife, daughter, daughter-in-law, son and son-in-law and friends by the ministers, bureaucrats, officials and development agents. They run NGOs in their pocket or houses. The get fund in the name of development, but grab money in their pockets without making any attempt to start the work of development. They show development on paper and submit the auditted account to the concerned ministries. Many NGOs have been black listed by the Government. Big NGOs get fund from the Government for development project, and they give them to small NGOs or development contractors on contract basis.

On the basis of above account the dilemma of development comes into picture that development is a welcomed by all, but criticised for increasing corruption.

Development is taken as Balanced Improvement, but it has Created Imbalance in Nature and Man

Development has been conceived by the planners as balanced improvement, but it has created imbalance in nature and between man. Industrial development, hydel project development, thermal power project development have undoubtedly brought positive result for the benefit of society, but the dilemma of development is that it has brought the problem of natural imbalance. Establishment of hydel projects have made several flora and fauna extinct. Industrial development have caused the problem of soil pollution, water pollution, air pollution and noise pollution. Thus, as a result of river valley projects and industrial development the problem of natural imbalance has come into existence. They have also created imbalance in nature-man-spirit complex. A number of abodes of spirits have also been made extinct for development projects. Development has created regional imbalance. Some regions have emerged as more developed and others as less developed. It has also caused societal imbalance. Some societies are developed and others are not developed. It has created rich man and the poor man.

The above given examples prove the dilemma of development that development is a balanced improvement, but it has created imbalanced in nature-man-spirit complex.

Development brings Improvement in Quality of Life, but it has affected the Quality of Life Adversely, too

Undoubtedly development has brought improvement in quality of life. The quality of life such as income, education, health, sanitation, safe drinking water, house, enjoyment of human rights etc. have been improved. But the dilemma associated with development is that it has affected the quality of life adversely, too.

The development has undoubtedly raised the per capita income. With increase in income, quality of life has also improved. But rise in income has caused attraction towards wine and liquor, luxurious items, drugs, unsafe sex relations, degradation of moral values, etc. Such practices have undoubtedly deteriorated the quality of life.

Education development has created unemployed youth. Unemployment has brought poverty, conflict, tension and other social pathological problems. Unemployed youth have been deviated from normal path. They have become victim of crime, drug addiction, frustration, moral degradation, murder, suicide, etc. Educational development has also increased the rate of dowry. Increase in rate of dowry is responsible for bride abuse, bride burning, bride murder, bride suicide etc.

Advanced medical technology is responsible for female foeticide. Use of pesticides and chemicals in agriculture is affecting our health adversely. Industrial wastes flowing in rivers have created the problem of safe drinking water. Smokes coming from chimney, motor vehicles, truck etc. have created the problem of air pollution in big cities. Big cities are also facing the problem of noise pollution. As a result of noise pollution, air pollution and water pollution, health, sanitation and drinking water have been affected adversely. The use of coolers, ACs, refrigerators etc. are also affecting the health adversely.

On the basis of above account the dilemma of development comes into light that development brings improvement in quality of life, but it has affected the quality of life adversely, too.

Development is Planned and Implemented for the Benefit of People, but it is Opposed by the People, too

There is no doubt that development is planned and implemented for the benefit of people. Undoubtedly people have been benefited by economic, educational, health and communicational developments. But there are certain developmental projects whose implementation is being opposed by the people through organized actions. People are opposing those development schemes which are expected to bring adverse effect on their economy, life, ecology and culture. People are opposing the cutting of forests ruthlessly for plywood furniture and paper industries, people are opposing the plantation of Eucalyptus trees, people are opposing the construction of koel-karo project, Tehari-Garhwal Project Suwarna Rekha project, Narmada Valley project, Sardar Sarovar Pariyojana etc. in organised way. The reason behind the opposition of these projects by the people is that installations of these project would bring devastating effect upon the life and culture of the people.

Thus, the dilemma of development appears before us is that development is planned and implemented for the benefit of the people, but it is opposed by the people, too.

Development is for Beneficiaries, but it is more Beneficial for Implementing Authorities

There is no doubt development is for the beneficiaries. The development programme is formulated and implemented keeping beneficiaries in view. Beneficiaries are most advantageous group in development. But the dilemma associated with development is that it is more beneficial for implementing authorities. Implementing agencies become wealthy and healthy by development projects and beneficiaries become more and more poor and unhealthy. Several reports dealing with the evaluation of beneficiaries of development schemes have reported clearly that beneficiaries are exploited, oppressed and deprived of development projects. Implementing authorities have shown names of beneficiaries on paper and have taken away the benefit in their own pockets. Poor, innocent and unconscious beneficiaries easily become the prey of clever implementing anthorities.

Beneficiaries remain more and more poor, but implementing authorities become more and more rich.

Development has divided People in FCs, OBCs, STs and SCs, but in each Category all are not Developed Equally

For the purpose of planning and implementing development schemes, our Government has divided us into categories like FCs, OBCs, SCs and STs. Each category includes several ethnic groups. All ethnic groups coming under a particular category have not been benefited equally by development programmes.

Under the category of forward castes, a number of such ethnic groups have been included who are economically not well. Again in an ethnic group, conditions of some families are well, while the conditions of remaining families are not good.

Under the category of OBCs, there are nearly 4000 castes. The economic conditions of all castes coming under this category are not equal. Some are highly developed, some are developing and some are still undeveloped. But as they all have been placed under the same category OBCs, the benefit of development is taken more by the developed ones.

Under the category SCs, there are nearly 700 communities. They are not equal on economic, educational and numerical scale. But they have been clubed together under the category SCs. The benefit of development programmes have been utilized more by the castes having numerical strength, better economic and educational condition. Some communities coming under this category are still undeveloped.

Under the category Scheduled Tribes we have nearly 533 communities. Numerical strength, economic condition and educational status of each community are not equal. But they have been jointed together under the category STs. In this situation, the community having more numerical strength, better economic condition and educational status have been able to develop more than the other communities of the some categories. Thus, the dilemma of development appears that it has created a number of categories but in each category all are not developed equally.

Development is Freedom from Bondage of all kinds, but it has Increased Institutional Bondage

There is no doubt that bondage leads to exploitation, oppression and deprivation. That is why Novel Laurete and famous international economist Amartya Sen has defined development as freedom from bondage of all kinds. But in the race of development people are becoming indebted to Self Help Groups (SHGs), private banks, national banks and other agencies. This has increased the bondedness of people. They are subjected to exploitation and oppression.

Development is Enjoyment of Human Rights, but has also led to Non-enjoyment of Human Rights

The process of development has created developed and undeveloped people. Through their employment bondage, multinational companies and other agencies are creating the situation of non-enjoyment of human rights.

3

Theories, Models and Approaches of Development

Numerous theories of development have been propounded by thinkers dealing with social and cultural development. For example, evolutionism, diffusionism, functionalism, structuralism, Marxism, capitalism, socialism, secularism, modernism and post-modernism. But here we are not concerned with these theories of social cultural development. Our concern here is theories related to planned development. In our country efforts regarding planned development are going on since the emergence of very First Five Years Plan. Now we have entered into Tenth Five Years Plan. In these plan periods theories regarding the development have been changed time to time after the evaluation of the performance of each plan. The various theories propounded by the builders of our nation for the planned development of people and nation are as follows:

1. Economic Theory of Development
2. Welfare Theory of Development
3. Industrialization Theory of development
4. Nationalization Theory of Development
5. Quality of Life Theory of Development
6. Human Right Theory of Development
7. Sustainable Theory of Development
8. LPG Theory of Development

1. Economic Theory of Development

When our country achieved freedom, the builders of our nation propounded the theory of strong economic growth for the development of people and the nation. It was thought that strong economic growth of the country will make the people developed and free from exploitation, oppression and deprivation. Under this development theory emphasis was given on agricultural development, irrigational development, cottage industry development, road and transport development. India was an agricultural country so greater emphasis was given on agricultural development in particular and over all economic development in general. Following this theory attempt was made in First Five Years Plan and Second Five Years Plan to raise the per capita income, Gross Domestic Product (GDP), Gross National Product (GNP) and National Income (NI). All development programmes were directed towards the achievement of economic growth. Economic Development Schemes are still going on.

2. Welfare Theory of Development

In order to bring the disadvantage and deprived sections of the country in the main stream of national development, welfare theory of development was propounded by our nation builders. India was declared as welfare state. Deprived sections were listed as Scheduled Castes and Scheduled Tribes to take up welfare measures for their development. Free education, hostel accomodation, lodging and fooding were provided to the children belong to STs and SCs. Reading materials and dresses were also given to them free of cost. It was thought that such type of welfare related scheme would help development of of the children of STs and SCs. Welfare Schemes are still going on for women, children and labourers of STs and SCs.

3. Industrialization Theory of Development

In the Second Plan Period industrialisation theory of development was propounded by the builders of our nation. Following this theory industries were set up in the areas having rich natural resources. Hilly and mountainous areas of our country were full of natural resources and mines. These areas were also backward in comparison to plain areas. Mostly tribal people were inhabiting in these areas. By following the theory of industrialisation industries were established in HEC Hatia, Ranchi, Jamshedpur Tata, Bokaro Steel City, (all in Jharkhand) Durgapur (West Bengal), Raurkela (Orissa), Bhilai (Chhattisgarh). Besides copper industries in Ghatshila and Jhikpani, Mica industries in Giridih and coal industries in Dhanbad were also established. It was thought that industrial development in these areas would help improve economy of the areas and the people. But establishment of these heavy industries in hilly mountainous and tribal areas could not improve the lot of the local people. Heavy industrial development also could not improve the backwardness of the areas. Rather it brought the problem of uprooting, rehabilitation, extinction, deprivation, exploitation and oppression.

In the Third Plan period, basic industries related to electricity, iron, petroleum and fuel got attention of our nation builders. Thus, industrialisation theory of development were established and followed in our country.

4. Nationalisation Theory of Development

In Fourth Plan period Nationalisation theory of development was propounded by our nation builder like late Smt. Indira Gandhi. Following this theory nationalisation of coal, mines, ores, industries, electricity production and supply, banking services, LIC, Hospitals, post office, Railways, Schools, Colleges etc., were made. This theory was pro workers and pro national economy, anti-exploitation, oppression and deprivation. This theory of development got support from every corner of the country. It brought happiness among the families of workers. Although we are again moving towards privatisation, but nationalisation theory of development is still continuing.

5. Quality of Life Theory of Development

This theory explains that development cannot only be measured in terms of economic growth like per capita income,

GDP, GNP, NI etc. But development is over all improvements in quality of life. Over all improvement in quality of life can be brought by income generation, education and knowledge, health, sanitation, drinking water, hygienic house, and standard of living. After Fourth Five Years Plan period, attempts are being made to improve the quality of life of people. A good number of schemes related to poverty alleviation and income generation schemes are being launched to bring the people above poverty line and to raise their standard of living. Educational schemes for large scale enrolment, stopping non-attendance and drop-outs have been launched to make people educated and to impart knowledge of various kinds among them. In order to remove malnutrition, anaemia and other kind of health problems, a good number of health schemes are being launched to maintain the health of expectant mother, nursing mother, infants, child and other people. Attempts are being made to provide safe drinking water to all people. Attempts are being made to provide sanitation to the people through the construction of community laterine, Sulabh laterine, safety tank laterine, drains, street and village connecting roads. Hygienic houses are being constructed for SCs, and STs. Thus, improvement in quality of life theory of development has attracted people more for participation in development programmes.

6. Human Rights Theory of Development

From mid-seventies development is being framed in terms of Human Rights. Human Rights are those rights which every human beings, male, female, children, aged, labours, local people and all people are expected to enjoy. Human rights deal with civil right, political right, economic right, social right and cultural right. Following this theory, Human Right Commissions have been set up at national level and state level. The Commission looks after the enjoyment of human rights. People are expected to approach these commissions if they are deprived from their human rights. Exploited, oppressed and deprived persons are free to seek justice through human rights commissions.

7. Sustainable Theory of Development

In order to save people, flora, fauna and ecology, sustainable theory of development has been propounded by our nation builders. This theory is pro-children, pro-women, pro-aged, pro-labour, pro-local people, pro-flora, pro-fauna, pro-environment and pro-ecology. This theory emphasises upon the balanced development. Important parts of this theory are as follows:

1. Rehabilitation of local people by land to land
2. Cultural rehabilitation of local people
3. Issuing identity card to rehabilitated families
4. Work guarantee to local people
5. Share of local people in development projects
6. Erection of hygienic houses for local people
7. Arrangement of schools, hospitals and market for local people
8. Soil conservation, water conservation, forest conservation and wild life conservation
9. Prevention of air pollution and sound pollution
10. Development of folk arts, sports and knowledge

This theory of development is for an heavy industrial projects and hydel projects.

8. LPG Theory of Development

Liberalisation, privatisation and globalisation are known as LPG Theory of development. This theory of development has been introduced in our country in 1990s and is still going on.

Following the liberalisation theory, Narasimha Rao Government had adopted liberal policies for foreign trades. Foreign countries were allowed to come to our countries for trade and business. The Government had lifted its ban on trade with several countries.

Following the privatisation theory Government privatised industries, hospitals, hotels, insurance, banks, airways, railways construction of roadways, waterways, bridge construction etc.

Following the Globalisation theory our Government entered into IT industries and services. Our villages became integrated with globe and the concept of global village developed. A good number of companies have come to our countries for their trade and business. They are providing jobs to our Engineers and management professionals. Economic globalisation, political globalisation and cultural globalisation of our country have taken place.

Models of Development

After Independence, our Government has introduced several models of development. Some models adopted by our Government for development of people are as follows:

1. Community Development Model
2. Agricultural Development Model
3. Irrigational Development Model
4. Electricity and Hydel Project Development Model
5. Cottage Industries Development Model
6. Heavy Industries Development Model
7. Mining Development Model
8. Transport Development Model
9. Communication Development Model
10. Education Development Model
11. Social Service Development Model
12. Women Development Model
13. Child Development Model
14. Labourers Development Model
15. Poverty Alleviation Development Model
16. Rural Development Model
17. Urban Development Model
18. Tribal Development Model
19. Scheduled Castes Development Model
20. OBCs Development Model
21. Rehabilitation Development Model

22. Sustainable Development Model
23. Environment Conservation Model of Development
24. Bio-gas Model of Development
25. Energy Model of Development
26. Regenerative Model of Development
27. Reproductive Model of Development
28. Health and Hygiene Model of Development
29. Sanitation Model of Development
30. Drinking water Model of Development
31. Disaster Management Model of Development
32. Privatisation Model of Development
33. Globalisation Model of Development
34. Nationalisation Model of Development
35. Social Development Model
36. Cultural Development Model
37. Handicraft Development Model
38. Political Development Model
39. Religious Development Model
40. Spiritual Model of Development
41. Social Security Model of Development.

As referred to earlier that development has many areas. That is why one model of development is not sufficient. The number of development models can be as much as the number of areas of development. Here attempt has been made to classify the development model on the basis of areas or fields of development. May be that we have forgotten certain areas and related models. Now let us discuss about these models one by one.

1. Community Development Model

Community Development model was started in our country on 2nd October, 1952 in 55 Blocks. This was an extensive programme for the development of village community formulated by Planning Commission. This model of development has sent encouraging messages among the people of village communities. Under this development model agriculture, transport, health, education, social welfare and agro-based industries were taken up.

2. *Agricultural Development Model*

Our country is an agricultural country. Soul of our country resides in villages where agriculture and agriculture related occupations are the only means of stay. The Government formulated the agriculture development model with the help agricultural scientists. Scientists of agricultural universities have extended full co-operation in agriculture development. Cultivation transplantation and harvesting by machines have come into existence. HYV seed, chemical fertilizers, pesticides, etc. were developed to increase the fertility of the soil. Scientific method of cultivation has emerged and undoubtedly agricultural fields have gone up. We are not only sufficient in food grains but we are also in a position to send food grains in foreign countries.

3. *Irrigation Development Model*

In order to increase the yield from agriculture, assured means of irrigation is a must. For this purpose, the Government has evolved canal system by making dam on rivers. In order to utilise the underground Source of water, the Government has adopted lift irrigation through Hiraboring. Bans (Bamboo) boring and Jaldhara yojna have also been planned and implemented. Million wells scheme have also been introduced along with diesel machine for the purpose of irrigation. Ponds have been dug out and their banks have been repaired time to time.

4. *Electricity and Hydel Project Development Model*

In order to provide electricity to railways, industries, rural areas, urban areas etc. hydel projects have been established. Through Bhakhra-Nangal Yojana, Rehanddam Yojana, Bhaisalotan Pariyojana, Tehri-Garhwal Pariyojana, Sardar

Sarovar Pariyojana, Damodar Valley Corporation etc. Electricity is also produced from coal and gas.

5. *Cottage Industries Development Model*

In order to facilitate the rural industries, Government has encouraged sugarcane mills, rice mills, wheat mills, pulse mills, oil mills, jute mills, dairy farm, fishery, horticulture, pottery, carpentry, blacksmithy etc.

6. *Heavy Industries Development Model*

In order to develop hills, mountainous and forested areas and also the people living there, our Government has evolved Heavy Industries model of development. These industries like HEC, Hatia, Ranchi, Jamshedpur, Bokaro (Jharkhand), Durgapur (West Bengal), Bhilai (Chhattishgarh), Raurkela (Orissa) are playing significant role in area development and development of the people.

7. *Mining Development Model*

Our country is rich in mines and ores, mica mines, copper mines, silver mines, gold mines, iron ores, tin mines, etc. are available in our country. Our Government has adopted model of development of these mines and people working there.

8. *Transport Development Model*

Transportation of goods and passengers is a must for development. With this view in mind our Government has adopted transport development model. For this purpose we have road transport, railway transport, water transport and air transport.

9. *Communication Development Model*

For the purpose of development communication of ideas, information, news and views are must. In order to facilitate communication our Government has adopted communication development models like radio, TV, cinema, telegraph, computer, internet, newspapers, post office etc.

10. Education Development Model

Education is very important means of development. It provides every individual to educate himself/ herself and enhance his/ her knowledge in the field of arts, science, commerce, technology, information technology etc. For this purpose our Government has introduced development models like primary, middle, Secondary and plus-2 schools, colleges, technical institutes, management institutes, research institutes, medical institutes, Yoga institutes, etc.

11. Social Service Development Model

Service of destitutes and disabled is like service offered to Gods and Goddesses. With this view in mind our Government has introduced social service development models through voluntary organisations, missionaries, non-Governmental organisations and International agenceis. They are running destitute homes, orphanage, oldage homes, mentally retarded homes, women rem and homes etc.

12. Women Development Model

Half of population of a nation is represented by women. But they have been victims of exploitation, oppression, deprivation, suicide, burning, poisoning and dowry death. In order to develop our women counterpart, Government has adopted models of development, like establishment of girl schools, women college, women technical institutes, girls hostels, women hostels, working women hostels. Seats have been reserved for them in educational institutions. Besides, seats have been reserved for them in panchayat, Panchayat Samiti and Zila Parisad. 33 per cent reservation in assemblies and parliament is under consideration. In order to prevent harassment of women on road and working sites, measures have been taken by government. Equal wages for women have been introduced. Maternity benefit schemes have been implemented. For health care of expectant women and breast feeding mothers, health schemes have been launched. Widows are allowed for remarriage and lead a prestigeous.

13. Children Development Model

Children are flowers of a nation. They are backbone of the nation. They are future hope of the nation. The development of a nation depends upon the development of the children. Considering this fact in mind our Government adopted children development model in the form of free education of all children up to age of 14 years. Free schooling, lodging, fooding, reading materials to children of STs and SCs, scholarship and coaching institutions for children of STs and SCs. Arrangement of book banks for them. For good health of mothers and children ICDS have been launched. Child labour has ben prevented. Bonded labour has been abolished. Abuse, exploitation, oppression and deprivation of children has been banned. Primary education to all children has been made a must, orphanage and destitute homes have been established. Children remand homes have also been established. DWACRA programme has also been launched.

14. Labourers Development Model

Labourers are the workforce of a country. They play very significant role in agriculture, mining, industrial and construction sectors. In order to save labourers from starvation, exploitation and oppression, Minimum Wages Act has been formulated, implemented and revised time to time by the Government. Bonded labour system has been abolished. Bonded labours were identified and released. They were covered under rehabilitation scheme. Debts of *mahajan* on the heads of labourers have been made free. Now labourers are free to sell their labour in open market at market value. Working hours, rest house, holidays and mode of payment have been determined. A labour inspector is posted in each block to look after the affairs of labourers. Women labourers have been allowed for equal payment for equal work. Sexual harassment of women labourers has been banned. Infants of women labourers are taken care of. 100 Days Work Guarantee Act has been implemented. Indira Awas Yojana has been introduced. Jawahar Rojgar Yojana has been implemented.

15. Poverty Alleviation Model

Poverty is regarded as major cause or hindrance towards development. In order to eradicate poverty the concept of Below Poverty Line (BPL) was developed. BPL families were brought under development schemes.Through developmental measures, they have been brought above poverty line BPL families have been made beneficiaries of development schemes like. JRY, Indira Awas Yojana, Gramin Rojgar Yojana, Gramin Sarak Yojana. Subsidy and loans have been distributed among them to raise income by small business.

16. Rural Development Model

In order to make villages developed, rural development model has been adopted by our Government. Village roads, streets, drains have been made *pucca*. Villages have been connected to block headquarters, railway stations, bus stand, market centres, hospital etc. through metalled road. Electricity is being provided to the villagers. Each village has primary Government Schools. In the radius of 5 kms, there is middle school and in the radius of 8 kms, there is high school. Some blocks have college, too. In each district , there is at least one college. Village have health subcentres, post office, chapakal, panchayat bhawan, community hall etc. Following this model of Rural Development by our Government, attempt is being made to make the village developed socially, culturally, economically and politically.

17. Urban Development Model

In our country, our Government has also adopted the model of urban development. Each town, city and metropolis are being developed through urban development planning. In the development of towns, cities and metropolis, emphasis is being paid on roads, streets drainages electricity, water supply, hospitals, railway, bus stand, transportation, post office, telegraph, internet, fax, colonies, markets, multistorey buildings, office, park, cinema hall, slum areas, gardens, etc.

18. Tribal Development Model

In order to make the tribals developed our Government has adopted tribal development models like Community Development Programme (CDP), Special Multipurpose Tribal Blocks (SMPTB), Tribal Development Agencies (TDA), Integrated Tribal Development Projects (ITDP), Primitive Tribal Groups (PTGs), Modified Area Development Approach (MADA) and Tribal Sub-plan (TSP). Through these models attempts are being made to make tribal developed socially, educationally, economically and politically. Scheduled Tribes Commission has been set up.

19. Scheduled Castes Development Model

Scheduled Castes are those castes who occupied low position in Hindu caste social stratification. They were deprived, exploited, oppressed and disadvantaged lot. Our President has listed them as Scheduled Castes to avail benefits of development and welfare schemes. Our Government has adopted prohibitive, protective and rehabilitative measures to make them developed. Untouchability has been abolished Atrocities on Harijan has been banned. Reservations in admission, jobs and MPs and MLAs has been provided. Residential schools have been set with free fooding, lodging and reading materials. Homestead land and agricultural land have been distributed. Colonies have been built for them. Loans and subsidy have been provided for income generation schemes. Human Rights Commission and Scheduled Castes Commission have been set up.

20. OBCs Development Model

OBCs are those castes who occupied middle position in Hindu caste hierarchy. They have been divided into Annexture-I and Annexture-II. OBCs coming under Annexture-II are more backward socially, economically and educationally. Our Government has evolved model for their development. Seats have been reserved for them in educational institutions and Government jobs. Some OBCs are more developed even better than many FCs. That is why the concept of creamy layer has been developed. OBCs coming under creamy layer are not entitled for enjoyment of reservational benefits. Previous OBCs having annual income of Rs. 1,00,000 were brought under creamy layer. But now it has been changed as Rs. 2,50,000 or taxable property or jobs as officers. Human Rights Commission and Backward Class Commission have been set up at national and state level.

21. Rehabilitation Development Model

Establishment of industrial and hydel project has brought the problem of uprooted families. For uprooted families rehabilitation development model has been evolved. Land for land, rehabilitation, compensation, service, and work guarantee schemes were followed for them. Now, there is proposal for cultural rehabilitation, too. Some share of profit is also to be distributed among them.

22. Sustainable Development Model

In order to make the impact of development sustainable, our Government has formulated a sustainable model of development. This model is pro-ecology, pro-health, pro-children, pro-women, pro-aged, pro-labourers, pro-environment, pro-wild life, pro-local people, pro-ecology and beneficial for uprooted families and rehabilitation beneficiaries. This model emphasises upon land by land rehabilitation along with cultural rehabilitation. Settlement of uprooted families and villages in healthy atmosphere along with all civil amenities, work opportunities and share in benefits.

23. *Environments Conservation Model of Development*

Everything that surrounds us is our environment. Only one people or generation does not have right over environment. All people and all generation have equal right to avail the benefit of environment. There fore, we must not bring imbalance in nature and environment. Keeping this fact in mind, our Government has evolved environment conservation development model. This model emphasises upon the conservation of soil, land, water, air, river, mountain, sea, ocean, lakes, ponds, flora, fauna and wild life.

24. *Bio-gas Development Model*

For the purpose of development, our Government has evolved a bio-gas model. This model emphasises upon the utilisation of biogases for the purpose of energy. Gobar-gas plants; Sulabh laterine plants, grass gas plants etc. are being established to utilise bio gases as an alternative source of energy development.

25. *Energy Development Model*

Besides K. Oil, diesel, petrol and uranium, our Government has evolved energy model of development. This model emphasises upon the use of solar energy, water energy and air energy for the purpose of development works.

26. *Regenerative Model of Development*

Our Government has also introduced regenerative model of development. This model emphasises upon the use of waste materials for the purpose of production. For example sugarcane waste, jute waste, rice chaff etc., are used for the production of papers. Garbages are used for wormi composte. Plastics are reused for making low grade plastics.

27. Reproductive Model of Development

In order to reproduce coal, iron, sugar, clothes, electricity, grains and varieties of things we need, our Government has introduced reproductive model of development.

28. Health and Hygiene Model of Development

Health is the greatest wealth. In order to maintain health in general and community health in particular, our Government has introduced health and hygiene model of development. Following this model, our Government has established health subcentres, health centres, *sadar* hospitals, medical colleges and hospital have been established. Allopathic, Homeopathic, Unani, Chinese and indigenous medical systems have been introduced. Drinking water and sanitation schemes have been launched. ICDS, DWACRA, Kishori Yojana. National Health and Nutrition Schemes, Minimum Need Programme, National Blindness Control Programme. National Filaria Control Programme, National AIDS Control Programme, National Iodine Defficiency Control Programme, Family Welfare Programme,National Kalajar Control Programme etc. have been formulated and implemented.

29. Sanitation Model of Development

The health of individual and community is directly related to environment cleanliness. Neat and clean house environment, street environment, drainage system cleanliness of chapakal, hand pumps, ponds etc. are essential for good health of community. Therefore, our Government has introduced sanitation model of development. Schemes of sanitary latrine, conversion of daily service latrine into sanitary latrine, separate latrin for community women, drain and street construction, awareness creation, conservation of hand pumps, well and ponds etc. have been formulated and implemented following this development model.

30. Drinking Water Development Model

Water is our life. In order to maintain good health of the community supply of fresh drinking water is a must. Therefore, our Government has employed drinking water development model. Under minimum needs programme, our state government provides, hand pumps in the locality of poor people, schools, hospitals, community halls, public places. UNICEF is also involved in supply of fresh drinking water to rural people. Our Central Government has launched a programme for supplying 40 litres of pure drinking water for each male and female. In plain area, one hand pump is to be established at a distance of 1.5 kms. In hilly area, source of water is to be developed at a height of 100 metres.

31. Disaster Management Model of Development

Flood, epidemic, earthquake, fire, desert, etc. are great disaster. These disasters cause harm to property and lives of people. In order to manage these disasters our Government has launched disaster management model of development. Under this development model, fire control, flood control, earth quake control; epidemic control etc. schemes have been formulated and introduced. Voluntary organisations and NGOs are also employed under this model.

32. Privatisation Model of Development

Development is not possible only by Government agencies. Therefore, private agencies are also introduced in development processes. Under this model of development, private sectors are invited and involved in development work such as education, health, industries, communication, transportation etc. Government keeps control over private sector.

33. Globalisation Model of Development

This model of development is recent one. Under the influence of unipolar word, our Government has introduced this model of development. This model has been adopted for cultural, economic and political development.

34. Nationalisation Model of Development

This model is opposite to privatisation model of development. Our country has mixed economy. Therefore, public sectors and private sectors both have been employed in development process of our country. Our Government has nationalised schools, colleges, banks, insurance, company, etc.

35. Social Development Model

Man is a social animal. He cannot live without society. For living together in society, social development is a must. There are some social virtues by which one can live peacefully in the society. Cooperation, sacrifice for other, respect, brotherhood, good conduct, etc. are social virtues. With the development of these virtues, one can lead life as a respectful member of the community.

36. Cultural Development Model

Man is a cultural being. He has evolved culture as a way of life. Culture is our social heritage. There is a urgent need for the preservation and development of our cultural heritage. Our monuments, cave dwellings, rockshelters, ancient palaces, temples, mountains, rivers, ponds, forests, arts, languages, dialects, drama, dance, music, songs knowledge etc. are our cultural heritage. In order to protect these heritage our Government has evolved cultural development and conservation models.

37. Handicraft Development Model

There are certain communities associated with weaving, basket making and other kinds of handicraft making. Our Government has introduced schemes for the development of their handicrafts. The Government provides them subsidy and loan to prepare handicrafts with modern design. Government also launches training programme to train them in preparing modern handicrafts. Government purchases their handicrafts for selling in market.

38. Political Development Model

Development of society and nation is also dependent upon the political institutions. The form of Government is responsible for development. The form of Government in our country is democracy and our Constitution is secular. In our country, democratic and secular model of development has been adopted under political development model.

39. Religious Development Model

Pity, alms giving, charity-forgiveness, tolerance, honesty, inter-communal brotherhood, peace etc. are necessary for religious development. Our Constitution is secular. Our Government has the policy of secularism. Religious missionaries are involved in development process. They are extending co-operation to the Government in development.

40. Spiritual Development Model

Man is spiritual being. Spirituality brings good health and peace of mind. That is why our Government has adopted spiritual development model. On the efforts of Government, Yoga institutes have been established. Yoga *shiwirs* are organised time to time on the initiative of the Government.

41. Social Security Model

In our country, joint family system used to provide social security to aged. But changes have occurred in this basic institution. As a result the problem of social security has emerged. In order to solve this problem, our Government has introduced social security model. Under this model, social security schemes like old age pension scheme, and establishment of old age home have been launched.

Approaches of Development

Approaches related to development are those ways by which development goals are achieved. After independence, our Government has followed a number of approaches to attain the objective of development. Some of approaches employed by our Government for the purpose of development are as follows:

1. Constitutional Approach
2. Political Approach
3. Administrative Approach
4. Welfare Approach
5. Commission Approach
6. Non-Governmental Approach
7. Missionary Approach
8. Anthropological Approach.

1. Constitutional Approach

Constitution is directive body of a nation. It directs the Government to take necessary steps for the development of weaker sections, backward areas and environment. Our Constitution is democratic and secular in character. It provides safeguards to weaker sections of our country. Our Constitution directs the Government to adopt following approaches for the development of weaker sections which includes Scheduled Castes, Scheduled Tribes, other backward classes, women, children and aged.

(i) Protective Approach

Our Constitution is committed to protect social, economic, political, religious, educational and environmental interests of the people through different articles and sub-articles.

(ii) Preventive/Prohibitive Approach

Our Constitution is committed to eradicate and prevent those social cultural practices which are exploitative, oppressive and deprivative for the people of our country through different articles and sub-articles.

(iii) Rehabilitative Approach

Our Constitution is also committed to rehabilitate the victims of attrocities, disaster, epidemic, earthquake, fire, flood, desert, etc.

2. *Political Approach*

For the purpose of development the political approach is a must. Our Government following the directive of the Constitution, formulates laws, makes amendment in the existing laws and implement those laws in the interest and development of the people. If the Government does not act according to the Constitution, people are free to take shelter of judiciary which is three tier arrangement, i.e. Lower Court, High Court and Supreme Court.

3. *Administrative Approach*

For the purpose of development, our Government has to adopt administrative approach. Administration includes implementation, monitoring , evaluation and follow up action related to development programmes. Administrative structure of development programmes includes ministry, secretaries, commissioners, deputy commissioners, directors, deputy directors, district magistrate, block development officers, VLWs, Panchayat sewaks etc. Through these administrative approaches our government formulates, implements, minitors, evaluates and follows up development programmes.

4. Welfare Approach

Our country is welfare state. Under welfare approach all-round development of weaker section is considered. The people belonging to weaker sections have been covered under different welfare schemes related to their educational, social, cultural, political and economic betterment. Under welfare schemes, Government meets cent-per-cent expenditure.

5. Commission Approach

In order to prevent exploitation, oppression and deprivation, our Government sets up different Commissions. The report of Commissions is discussed in the parliament. On the basis of reports and discussions, new laws are formulated or amendment in existing laws is made. Guilty is also punished on the reports of the Commission. Our Government has established Laws Commission, Human Rights Commission, Women Commission, Minority Commission, Scheduled Castes Commission, Scheduled Tribes Commission, OBC Commission, Farmers Commission, Labour Commission and Knowledge Commission. These Commissions function to prevent deprivation of the concerned sections. They also report about the performance of different welfare and development programmes.

6. Non-Government Approach

Government alone is not responsible for the betterment and development of the weaker sections. Non-Governmental agencies are also involved in the process of development by the Government. Under non- Governmental approach National or International level NGOs, voluntary organisations, social workers, social activists and environmentalists etc. are involved in the development process of weaker sections and areas.

7. Missionary Approach

Our Government has adopted missionaries approach also for the development of weaker sections. Missionaries are committed for the development of weaker sections. They render their services for educational, health, cultural, social and economic development of the people. That is why our Government has included them in the development process of the weaker sections. Christian missionaries, Ramkrishna Mission, Arya Samaj, and other religious missionaries are involved by our Government in the development processes.

8. Anthropological Approach

The government has also followed anthropological approach for the development of the weaker sections particularly tribals. Anthropological knowledge has been employed in following protective, prohibitive and rehabilitative measures. Anthropological knowledge has also been brought in action for the development of weaker sections. Anthropologists have been employed on different administrative post for development purpose. They have also been provided opportunity to work in different commissions set up by the Government to review the development programmes.

4

Cultural Indicators of Development

From the complex point of view, culture is a complex whole which includes moral, art, belief, attitude, knowledge, value and all other capabilities acquired by man as a member of society. Thus, culture is a complex whole. It is social heritage. It is acquired as member of society. It is not inherited genetically. From this point of view, development value, knowledge, belief attitude and morality are there in culture. A society accepts or rejects developmental input according their knowledge, belief, attitude, morality and value. These cultural traits are not stable, rather they are changing from time to time.

From the functional point of view, culture is mechanism, means, or instrument for the satisfaction of human needs. Human needs are numerous. The present needs are different from past and future needs. The first task of culture is satisfaction of survival needs. It also sets up conflict-resolving, consensus-building and problem solving mechanism

Development is meant as over all improvement in quality of life. The quality of life depends upon conceptualizing human need structure in new perspective and making adequate provision for it. Therefore, cultural indicators of development are related to satisfaction of human needs. It is very difficult to offer an inventory of needs that would respond adequately to diverse cultural demands and aspirations, yet some common needs indicating cultural indicators of development can be listed as follows:

1. Survival Needs
2. Social Needs

3. Welfare Needs
4. Psychic Needs
5. Cultural Needs
6. Adoptive Needs
7. Progressive Needs.

1. Survival Needs

Development is related to fulfilment of survival needs. Satisfaction of survival need is one of cultural indicators of development. The society or nation the survival needs of whose members are satisfied comes under the category of developed society or nation. The satisfaction of survival needs includes provisions for food, shelter, gainful employment, preventive and curative medicine and protection of life and property. For the purpose of development, satisfaction of survival needs requires basic alterations in distributive value and reward system. It also requires a curl on individual consumption and on extension and enrichment of social services.

2. Social Needs

Development has social dimensions. Therefore, satisfaction of social needs has also been included under one of indicators of development. Satisfaction of social needs involves promotion of community-ness, social cohesion, creation of effective conflict resolving and consensus building mechanism and evolving and enforcing norms of social discipline. The social needs assume growth of social consciousness, restructuring status and role and a major human engineering effort with ascent on cooperation, consensus and discipline.

3. Welfare Needs

For development satisfaction of welfare need is also a must. That is why the satisfaction of welfare needs have been included in one of cultural indicators of development. Satisfaction of welfare needs includes measures that offer a fair deal to weak, disabled, handicapped and vulnerable. It is related to protection of weaker sections so that they could develop themselves and contribute in development of the nation.

4. Psychic Needs

Development is related to satisfaction of Psychic needs. For development, satisfaction of psychic needs is also a must. Without satisfaction of psychic needs, one cannot participate in development process. The satisfaction of psychic needs include provisions of personal freedom and privacy leisure and its creative utilization and equal opportunity for advancement and general development.

5. Satisfaction of Cultural Needs

There are two aspects of culture. One is material aspects and other is non-material aspects. For development, satisfaction of material and non-material cultural traits is a must. Therefore, both material cultural needs and non-material cultural needs must be satisfied for development. Material possession and advancement are useless if they do not provide peace and harmony in the society. Therefore, material development and non-material development must not be in isolation. They are integrated and supplement to each other. Material advancement must be as per value and morality of the society. It must ensure equality, peace, harmony and interactions in the society.

6. Adoptive Needs

Culture is adoptive and accumulative. It is not static rather dynamic. For development, satisfaction of adoptive need is a must. Satisfaction of adoptive needs, requires mechanisms for scanning the social, cultural, psychological and physical environments as well as identifying and effecting modifications necessiated by changes in these environments.

7. Progressive Needs

Culture is progressive. It makes progress. It is true that incourse of progress certain cultural traits get degerated. But it accepts modifications and makes progress. Therefore, satisfaction of progressive needs has been included as one of indicators of development. Satisfaction of progressive needs requires sharpening of problem anticipating and problem solving capabilities. It also requires growth of scientific and developmental researches and development of human engineering skills. On the basis of above discussions, it can be said that there are cultural indicators of development. Cultural development can be assessed and analysed on the basis of these indicators such as satisfaction of survival needs, social needs, welfare needs, psychic needs, cultural needs, adoptive needs and progressive needs.

5

Social-Cultural Factors Supporting Development

Man is a social and progressive animal since time immemorial. Man has started making inventions from Neolithic period. Inventions and discoveries are still going on in our society. These inventions have made our life more and more easy. Although man is progressive and developing animal, but the pace of development has not been equal for all individuals and societies. There are individuals and societies less developed or undeveloped in comparison to other individuals and society. Such differences in development are due to variations in environments as well exploitation and deprivation of resourceless by resourceful persons. It is beyond doubt that development is a social cultural in nature. Without the support of social cultural factors, development is not possible. Chief social-cultural factors supporting development are as follows:

1. Family

Family is smallest unit of society. It is centre of social, cultural, economic, political, religious and educational activities. An individual is born in a family. There is father and mother to welcome his birth. He gets family name, lineage and *Gotra*. Thus, individual gets recognition in the society after his birth. If family does not take care of infants he/she cannot survive. The family has concept of disease and medicine for pregnant women and infant. In each family, there is concept of health and hygiene. Through the process of socialisation knowledge about disease and treatment is transmitted from one generation to another generation. Family develops good and healthy habits in children. For own development and participation in development of society and nation, good habit and good health is a must. Through socialisation process, social virtues like pity, tolerance, non-violence, co-operation, excuse, etc. are developed in the children. These social virtues help him much in the development of good personality and good member of the family, society and nation. Family is regarded as first school and mother is regarded as first teacher in our value system.

For the development of family, cooperation, confidence and faith among family members are necessary. Parents invest their maximum part of income on the health and education of children. Good relation between father and mother, father-son father-daughter, mother-son and mother-daughter is a must for the development of family. Children makes development because their parents make efforts for their development. Only effort of parents is not sufficient cooperation of children is also essential for development. Many parents become disheartened on the behaviour of children. Their children do not want to read in spite of all efforts of the parents. Sometimes quarrel between parents becomes the barrier of development. Sometimes joint family system come in the way of development as resistance.

2. Neighbourhood

Whether a family lives in rural area or urban area, there are neighbours with whom it lives. The development of a family also depends upon the environment of neighbourhood. A good environment of neighbourhood is considered beneficial for development. Bad environment of neighbourhood affects the development adversely. Good neighbours have control over the behaviour of children of neighbouring families. When they observe a child following bad habit or doing misbehaviour with community member, they prohibit the child and inform their parents about the misbehaviour and bad habit. Among good neighbours, there is competition in the education of children. Each family of neighbourhood wants to provide good health to their children. They get children admitted in good schools and colleges. Children when visit in neighbouring families with their age group, they are encouraged for good education and career. Neighbours also seek advice from each other regarding the education and career of the children. Good neighbour makes life heaven. They cooperate in joy and sorrow. They cooperate in making outside home environment neat and clean. Bad neighbour makes the life hell. They bring miseries and troubles of various kinds. They also create resistance in development. When a boy of neighbourhood becomes engineer, doctor, IAS, IPS, professor, lawyer etc. he or she becomes guide for other children of the neighbourhood. He/she as well his/her parents also encourage the children to study more and more and to select good career.

3. Kin groups

Kin groups play significant role in development. It has been observed in the society that kin groups extend help in the development of each other. They extend cooperation and help during miseries. If one relative occupies a good post after his good education, he/she becomes a matter of daily discussion among his/her relatives. Relatives pay a visit to seek advice regarding the education and career of their children. He/She acts as career counsellor for relatives. Relatives also extend help in meeting expenditure on education, health, erecting house, marriage, etc. It is also true that instances of kin kitting and kin conflict also appear in the society. But it is also true that kins play valuable role in the development. They provide good career counselling and encourage for developmental works. People take help of kins in development. It is found that an appointing authority help his relatives in getting them appointed. Instances of brotherism and nephewism come in to existence in appointment. Kins men are found working together and living together in urban and industrial colonies.

4. Caste

Caste is an important social factor. Its role in development cannot be neglected. In our country castes have been categorized politically as forward castes, or general castes, other Backward castes, Scheduled Castes and Scheduled Tribes. Forward castes are more developed socially, educationally, politically and economically. For OBCs, STs and SCs, reservation policies have been adopted as per provisions laid down in our secular and democratic Constitution. Politically caste is an important factor of development. For the purpose of development and to bring all people in the mainstream of development, castes have been categorised and reservation policies have been adopted.

Caste is also criticized for casteism and caste riots. Casteism helps in development of caste members. It has been observed that appointing authority favours members of own castes at the time of appointment. One also finds colonies in urban areas in the name of castes.

5. Village

Village community is also a social factor which plays significant role in development. Families living in Villages have we feeling. Educated people of the village become education and career counsellor of children of village community. They extend full help in educational development of children of the village. If a person becomes appointing authority, he helps his villagers beyond caste in getting them employed. Villageism prevails at the time of appoint and development works.

6. School and College

Education is main means of development. It provides occasion to select career and modern occupation. It also makes people aware of citizen right, human right and development right. In school, a boy/girl enters into a new environment. Here, there is no father and mother to take care of them. Here he/she finds himself/herself amidst teacher, students, some clerks and staff. The environment of school is quite different from the environment of home. Here, he/she has to maintain discipline. He/she is taught about discipline, honesty, brotherhood, tolerance, non-violence, love, pity, science, technology, inventions, etc. Educational knowledge goes on increasing as one passes from one class to another. After primary, middle and secondary education, one enters into plus-2 school or inter colleges. From there, he/she enters into general or technical colleges. He/she also enters into university education for post graduate degrees. After attaining degree one enters into career and service. School, colleges, and university provides the occasion of knowledge enhancement and career building which are quite essential for development.

7. Friend Group

In colleges and universities, friend group is established. All students forming a friend group seat together in class, live together in hostel, eat together in mess and discuss problem related to their study and career. Friend group is beneficial for study and career counselling.

8. Play Group

Students are youth. They form play team or group for playing games and sports. In order to win match team spirit is a must. Without team spirit, one cannot think about success in matches. Play group is also helpful for development. Play group is beneficial for career counselling in games and sports.

9. Occupation Group

Occupation group also plays significant role in development. People working together at a place or in an institution extend help and cooperation to each other during trouble. They are together in enjoy and sorrow. People seek advice regarding education and career of children; treatment of disease, promotion, house construction, saving; investment, marriage of children etc. On the advice of occupation group, people take steps related to development of children and other things.

10. Marriage

Marriage is association between two individuals of opposite sex. It gives rise to family and places responsibility of reproduction and care taking of children on the couple. Now-a-day marriage is placing significant role in development. Previously, women were prohibited for enrolment in schools, and colleges. They were kept only in the boundary of house as housewife. But today education of women is an important factor in settlement of marriage. Now women are also conscious of their education and career. They are found working in every fields along with their male counterparts. Now-a-days, marriage between educated and careerist boy and girl is preferred. This is very positive for the educational development and career of woman.

11. Religion

Religion is an integral part of our society and culture. In religion, we find belief related to superhuman powers and rituals in order to please them for fulfilment of desired goal. Development is also related to fulfilment of desire goal such as education, career, occupation, good health, good wealth, good harvest and good progeny. In order to achieve these developmental traits, people go in the shelter of super human powers as per their belief system. They make special promise to offer special worship after attaining education, career, occupation wealth, good harvest, good progeny and good health in the family. People please their Gods and Goddesses for peace and prosperity in the family. Religion also teaches us lesson about truth, honesty, non-violence, brotherhood, social cohesion, peace and prosperity. People offer special worship to their Gods and Goddesses whenever there is educational, social, economic, political and health achievement in the family. People do good deeds, make *dan* (charity) and *punya* (virtuous) for attaining *moksha* (salvation) in life after death. Thus, our religion is beneficial for.development.

12. Tradition

Culture passes on from one generation to another through tradition. Culture is not biological heritage. It is social heritage. No one comes with his/her culture from the womb of mother. An individual learns culture in his society through the process of socialization. Our culture is progressive, accumulative and developing since time immemorial through inventions, discoveries and knowledge. Our inventions, discoveries and knowledge pass on through tradition. It is generally said that traditions are not broken, but something can be added to them. It means that traditions are not static, rather they are changing. Changes are accommodated in traditions. The study of traditions tell us about the development of knowledge, science, technology, medicine, law, administration, government, state etc. Thus, our traditions are helpful in accommodating development for the benefit of people.

13. Customs

Customs are above law. They are obeyed by us because they are related to our personality development, social development, health development, economic development and religious gains. From birth to death and religious festivals, we have to obey a number of customs. Customs is obeyed at societal level. It is related to societal development.

14. Faith

Faith plays significant role in development. The path of development is not easy. It is full of trouble and unsuccess. Trouble and unsuccess deviate us from the path of development. But our faith provides us strength of moving forward on the path of development. Positive thinking and faith provide us with developmental success. In order to make our faith and will power strong, we go to the shelter of our God and Goddesses. We seek blessing from our Gods and Goddesses. Their blessings provide us with developmental success related to education, occupation, wealth, health, progeny, etc.

15. Ritual

Faith is thought and ritual is action. The ritualistic action makes our thought more and more strong. For achieving developmental success, we perform a number of rituals before our Gods and Goddesses. Before erecting house, multistorey flats, school, colleges, hospitals, administrative buildings, making bridge etc. the worship of land is done. The worship of land makes our faith strong regarding the success of the project. When the project is over, inaugural ritual is performed. We perform the worship of mother Parvati, Lakshmi and Saraswati for progenys health, wealth and education in the family. Satyanarayan Swami Katha is heard for fulfilment of good health, wealth, education and progeny. For attaining progeny, *Harivansh Purana* is heard. For wealth mother Goddess Lakshmi is worshipped. For pleasing death planet, and good health *Mahamrityunjaya Jap* is done. For attaining peace on earth, *Yagya* is performed by Rishi-Munis and Sadhus. Thus, ritual plays significant role in development.

16. Festivals

Celebration of festivals develops social cohesion, communal harmony, brotherhood. Celebration of festivals is related to social, cultural, economic and religious development. Celebration of festivals requires co-operation and co-operation brings development of relation in the society among different segments. Festivals reveals victory of good over bad, and God over devils. Victory of good and God brings development in the society.

17. Art

Art is our social and cultural heritage. It has become an important means of career, too. We depict our Gods, Goddesses, temples, monuments, love, affection, feeling etc. through art. Art forms are many. Art exhibition brings people together. Art performers also get prize. Thus, art and performing art play valuable role in development of sentiment, feeling, affection, love etc. Exhibition of arts also plays valuable role in development.

18. Music and Songs

Music and songs are also helpful in understanding development. Music and songs also reveal their own development in different periods. Besides, they tell us about many evil practices and attempt of the Government in eradicating them. They also make people aware of the developmental programmes and their impact. Many poets compose songs on development issues and make people aware by singing them along with playing musical instruments. Music and song programmes of Radio, Television and Cinema also relay development related songs followed by music.

19. Dance

Dance is performed individually or in group. We have now various kinds of dances traditional dance, local dance, regional dance, national dance and modern dances. All these dances have their own developmental features. Besides, they are helpful in development. Development related messages are conveyed to the people through folk dances or through modern dances. People easily catch those news.

20. Folk Drama

Folk drama is significant cultural items of the folk society. Folk drama communicates easily to members of the folk society about social evils, attempts of the Government to eradicate them and development programmes going on for the people. Folk drama makes people development conscious.

6

Impediments to Development

There is no doubt that development is a term which is good and pleasant to hear. Everybody, every family and every community makes attempt for development. But development is not such a thing which can be achieved easily. It needs patience, time, planning, policies, resources, implementation, monitoring, evaluation and follow-up action. So, development is a time taking complex phenomenon. It has also to face the situation of impediment. Undoubtedly there are factors supporting the development, but it is also true that there are factors which create impediment to development. Impedimental factors of development can be discussed under the following headings for the purpose of study:

1. Cultural Impediment
2. Social Impediment
3. Economic Impediment
4. Political Impediment
5. Religious Impediment
6. Psychological Impediment
7. Natural Impediment
8. Administrative Impediment
9. Corruptional Impediment
10. Educational Impediment
11. Health Impediment
12. Terroristic Impediment
13. Ecological Impediment
14. Communicational Impediment
15. Naxal Impediment.

I. Cultural Impediment

Culture is a complex whole which includes attitude, belief, custom, knowledge habit, and all other things acquired by man as a member of society. Our attitude, belief, knowledge, habit, custom, tradition etc. create impediment to development. Let us discuss about these cultural traits as impediments to development one by one.

1. Attitude

Our tradition and custom play significant role in formation of attitude. In our traditional society, attitude towards education of girl and work out-side home is not positive. Due to this attitude, women in our society have to suffer much. They have been deprived of education and modern job opportunities on account of this cultural attitude.

Attitude of people towards hand pump for safe drinking water is also not positive due to use of leather made washer. Leather made washer makes the water impure.

Attitude of people towards untouchability is also not positive. People maintain distance from them.

Similarly attitude of the people towards laterine in bedroom is also not positive in rural areas.

These attitudes of the people are now under change, but they have created impediments to development of concerned people.

2. Belief

Our belief is an integral part of our cultural system. Our various beliefs compell us not to follow progressive path and creates hindrance in development. Being guided by our belief, we have holy bath in dirty water of ponds and rivers, we drink water of dirty well, we do not go to doctor and instead go to *shamans*, for treatment, we abuse women in the name of witchcraft, we treat women unclean in menstrual period, we stop journey considering some days, time or event inauspicious etc. Such types of our belief create impediments toour development of various kinds.

3. *Ritual*

Rituals are action part of the belief. Ritual include *puja-path* or worship to Gods, Goddesses and spirits. The performance of rituals requires materials like sunfried rice, ginger, mustard seed, oat, black *Til*, coconut, paddy, mango leaves, *kusha* leaves, *palas* leaves, *karam* leaves, *sal* leaves, curd, *ghee*, milk loafsugar, *dhup*, *agarbatti* and sacrificial animals. Their performance needs specialists and time. The performance of ritual is followed by exchange of invitation, visit, food feast gift and presentation. In this way, performance of ritual is time taking as well costly affairs. There is no harm in the performance of rituals, but harm is when rituals creates impediment to development. It is observed that for the performance of rituals people go to the *mahajan* to take loan on heavy interest. They become victim of indebtedness and bonded labour. The *mahajan* begins to exploit them in a number of ways. Under bonded condition, they could not remain free. They are deprived of their human rights. Ganerally poor people are more followers of ritualistic traditions. That is why they become more and more victim of exploitation. As a result, they remained undeveloped. They do not spend on health, treatment of disease and education. They are misguided in the name of rituals.

4. *Habit*

Habit is also one of cultural elements. It is also known as second nature. It is formed during childhood and continues till death. Some habits are good and some habits are bad. Good habits are taken as good by the society, but bad habit becomes the matter of criticism in the society. Although in every culture, there is socialisation process to build good habit in individual during childhood, but an individual learns some bad habit too. Bad habit develops due to ignorance and not proper care. Bad habits are taken as detrimental for the development of own health as well as the health of the community. Some in individuals do not put soil on night soil after defecation. This bad habit spreads dirt in the environment which ultimately affects the health of the community adversely. Some individuals do not wash hands and feet after defecation, they do not wash mouth properly, they do not cut nails, they do not wash hands and mouth before and after eating properly, they do not chew food properly. They do not drink water sufficient for the body, they sleep in late night and do not rise in the morning, they wear dirty clothes, they do not have regular bath, they do not keep house neat and clean, etc. These are bad habits which cause damage to the health of the individuals concerned as well as the health of the community where such persons with bad habits live.

Habit also compells us not to accept schemes of development. For example, an evaluation study of community laterine scheme for rural women revealed that the scheme could not get success due to habit of women to use field. Women of the concerned community used to meet each other, have gossip, fun discussions on family matter and events etc. at the time of defecation in the open field. Community laterine scheme was not suitable for their cultural habit. That is why they were not attracted towards the laterines constructed for their use. Pigs were using those laterines for their stay.

5. *Customs*

Customs are cultural elements. They are above law and have positive sanction of the concerned community. But some customs are detrimental to the development of health, community and society. For example, custom of child marriage, custom of prohibition on widow remarriage, custom of dowry, custom of sale and purchase of girls, custom of not giving share to girl in paternal property, custom of female infanticide etc. These customs are detrimental to the educational, health and social development.

6. *Tradition*

Cultural traits transmitted orally from one generation to the next are termed as tradition. Traditions are cultural elements. It is said that traditions are our heritage. They are not broken, rather they are maintained. Our cultural traditions also create impediment to development. If some attempts are made to provide intervention in the existing tradition of the community through development schemes, people have reacted negatively, for example, Birhor tribe. Birhor is a little known tribe in the state of Jharkhand. It has two sections *Jaghi* (Settled) and *Uthalu* (Wandering). Attempt was made by the Central as well as State Governments to make the *Uthalu* Birhor settled by erecting a colony for their stay with drinking water and agricultural facilities. But *Uthalu* Birhor abandoned the colony because it was not supporting their cultural traditions.

Similarly, in tribal and rural societies, people are still not giving the first milk of mother to the newly delivered child. Instead, they put first milk on earth to satisfy evil spirits. This tradition is harmful for the physical development of the infant.

7. Knowledge

Knowledge is one of cultural elements of our society. Knowledge is our heritage and is transmitted from generation to generation. Society has its Traditional knowledge of agriculture, fishing, animal husbandry, erecting house, health, hygiene, basketry, pottery, blacksmithy, coppersmithy, bronzesmithy, carpentry, health, desease and treatment. Traditional experts of knowledge provide their services orally which continues from one generation to the next. They train their sons in their specialisation. But knowledge must be scientific and complete. Unscientific and incomplete knowledge creates impediment to development.

8. Morality

Every society emphasises upon its moral codes of behaviour. It emphasises upon the character building. Morality and character are significant cultural element of a society. They are our cultural heritage. But competition to earn more and more money and material gains has devalued morality. For earning money characterlessness has appeared in society. This is very harmful for cultural development. People are adopting immorality and characterlessness for earning money. They come as impediment to moral and character development.

9. Art

Art is an important cultural element of a society. It includes making artistic design and performing arts. An artist designs his feelings, emotions, expression, opinion etc. through his/her drawing and painting. Similarly an artist performs art through dance, drama, music and song. But now -a-days money has come in the way of art. Now-a-days folk art and folkartists are being neglected. They do not have money to compete with modern artists. A number of folk arts have either vanished or are in the way of vanishing. Their gradual disappearance is really a cultural loss. Art competition has created impediment to development of folk arts.

II. Social Impediment

In society there are a number of social practices. Social practice places emphasis upon observance of a number of taboos and restrictions. The observance of taboos and restrictions create impediment to social development. For the purpose of study we can divide social impediments to development under the following headings.

1. Nuclear Family

Family plays significant role in social, cultural and educational development of children. Family is the primary school with mother as its teacher. The mother as teacher teaches children about social norms, cultural values and habit building. Father provides control and guidance in character building of the child. But social, cultural and educational development of children take place only in those families whose parents are conscious and aware of development. It is also true that peace and prosperity come only in those families where relation between parents and between parents and children are sweet and harmonial. In contrast to it the families whose relation between husband and wife is conflicting, proper development of children does not take place. In families where husband is drunkard and beats his wife and children frequently, the development of child and peace and prosperity in family can not be expected. Thus, tensile and conflicting relation between husband and wife and parents and children create impediment to development of family in general and children in particular.

2. *Joint Family*

India is an agricultural country. For agricultural practices joint family system is most suitable. It is characterised by common head, common house and common hearth. Besides common paternal or purchased property. But joint family is creating impediment to educational development of children. In this family all children are treated equally whether they are good in study or not. Children performing well results in examination do not find fund from the common head for study. Due to this even meritorious child has to leave school or college and join hands in agriculture. If not, conflict starts between families over the education of children. Conflicting situation retards the progress of the family. Division of joint families sometimes leads to murder of its own family members. Thus, joint family system on many occasions appears impedimental to development.

3. *Neighbourhood*

Neighbourhood also plays significant role in social development. A family is fortunate enough whose environment of neighbourhood is good. Good neighbours help each other in social and educational development. Children learn about social virtues, norms, values and character from the neighbourhood. A developed, well behaved and prosperous neighbour becomes a model of development for the families of neighbourhood. A good neighbour encourages his neighbouring families to maintain good and harmonial relation and to educate children in proper manner. Contrary to this a family is misfortunous whose environment of neighbourhood is bad. Drunkard, theft, dacoit and other antisocial elements living in neighbourhood spoil the environment of development of other families, too. Thus, bad neighbourhood is impeditive to development.

Bad neighbours use to quarrel with each other. They do not pay proper attention towards their development. They are involved antisocial activities like theft, dacoity, loot, burning, rape, murder etc. These anti-social activities in neighbourhood have negative bearing on developmental processes.

4. *Caste System*

In our society caste system is practiced. This system creates impediment to development of those castes who occupy lower position in caste hierarchy. The castes occupying lower position have been deprived of their rights. As a result, they have remained undeveloped, and resourceless. But now they have been listed as Scheduled Castes. Our Government is trying its best to make these communities developed socially, economically, politically and educationally.

5. *Casteism*

Casteism is different from caste system. In caste system, we find hierarchy. But in casteism we find loyalty of a person towards his/her own caste at the cost of other castes. Under casteism, members belonging to own caste are favoured and uplifted, while members belonging to other castes are disfavoured and supressed. Casteism is beneficial for own caste but detrimental for the other castes. Due to casteism prevalent in selection process of jobs, a number disabled candidates are selected and able candidates are rejected. Able candidates are subjected to frustration and disable candidates get opportunities to develop more and more. In this way, casteism creates impediments to development of able persons. It retards their development. Some of them become frustrated and are involved in antisocial activities, while some others are compelled to commit suicide.

6. Caste Riots

Now-a-days caste riots have also appeared. In caste riots members of one caste become enemies of other caste. Involved castes take life and property of each other. In this situation, members of both castes show loyalties to their own castes. They cause harm to each other as far as possible. In the situation of caste riots, children, women and aged are worst affected. Such riots creates impediment to development of all kind. In this situation, a man becomes animal. Like animal he forgets about development.

7. Village Factions and Feuds

Intra-village and inter-village conflict creates impediment to development. It causes loss of life, property and money. Energy and money have to be spent on police, thana and court. Intra-village and inter-village factions and feuds take place due to premarital or extra-marital sex relation, rape of a woman, theft of property and cattle, over the issue of land, river bank, fishing pond etc. Sometimes, situation becomes worst. Incidence of bloodshed and murder also take place. People have to spend on thana-police, court, lawyer, travelling, witness, etc. Peace and prosperity is snatched away. Money and energy are spent not on development, but on other issues related to bloodshed and murder. In this situation it is very difficult to think about development. Thus, village factions and feuds are impedimental to development.

8. Discrimination against Women

Discriminatory behaviour found in society against women is impedimental to their development. They are discriminated in rearing and feeding process which is detrimental to their physical developement. They are discriminated in education due to which they remained unenrolled in school and they become illiterate for ever. They are kept within the boundary of the house which retards their social development. They are not given share in paternal property or property earned by their parents which retards their economic development. They are not allowed to participate in traditional village panchayat which is detrimental to their political development. They are allowed to serve food after serving all members. On many occasions they remain hungry due to shortage of food. This is harmful for their health development. On becoming widow they are not allowed for remarriage and to grab the property of their husband they are abused as witch. Thus, discriminatory behaviour prevalent in society is harmful and act as barrier for social, educational, economic, political and health development.

III. Economic Impediment

For the purpose of development resources and funds are required. In absence of resources and funds, development is not possible. Resources are not equally distributed. Therefore, every individual or family does not get equal and proper opportunity of development. In our society, economic impediments to development can be discussed under the following headings:

1. Poverty

Poverty is a condition in which an individual or a family is unable to feed himself or his family both ends. Such families come under below poverty line and they are designated as poor or BPL families. These families have to struggle hard to main their existence. For these families bread for survival is more important than development. Majority of daily wage earners come from these families. Being poor, they are exploited by resourceful persons or their daily employers in a number of ways. Poverty makes people apathetic towards development. In this way poverty is impeditive to development.

2. *Unemployment*

Unemployment is also one of economic impediments to development. The situation of unemployment makes a family more and more poor. It makes the condition of a family more and more deplorable and pityable. Regular income from employment brings happiness in a family. But regular situation of unemployment brings miseries in family. Now-a-days rural youth and urban youth both are facing unemployment problem. Unemployed youth are turning towards antisocial activities like theft, dacoity, loot, snatching, rape, murder, etc. They are also becoming frustrated. Some of them become victim of depression and end their lives by committing suicide. In this way, unemployment is impeditive to social, cultural, educational, economic, physical and personality development.

3. *Landlessness*

In our society distribution of resources particularly land is not equal. Some are big land owners, some are small land owners, some are marginal land owners and some are landless. For landless people, their physique, health and hands are only means to earn livelihood. In other words, for them working as daily labour and earning wages are the only means of stay. As daily labourers, they have to face the situation of unemployment. They do not get work regularly round the year. During peak agricultural season, they get work but in off agricultural season they remain unemployed. As a result of this they are not in a position to have any kind of saving. Rather they remain struggling to maintain their existence on the earth. Being resourceless, they are exploited in payment of wages. They do not get payment as per Minimum Wages Act. Females get less payment than males for same type and hours of work. Exploitation prevalent in mode of payment forces them to become more and more poor day by day. In this situation of struggle for existance landless family even cannot imagine for development of various kind. Thus, landlessness is one of economic impediments to development.

4. Lack of Irrigation Facilities

In our country peasants are dependent upon rainfall. There is good harvest if rainfall is good. But when there is no rainfall, there is no harvest. People have to face the situation of drought and femine. In absence of irrigation facilities, peasants remain only owners of land. They do not become producer of yields from the land. Continuous situation of no rainfall or less rainfall, makes a land owning families poorer and poorer day by day. They have to struggle hard to maintain their families. In this situation, one cannot think development. In this situation one can think only to protect himself from the extinction. In this way lack of irrigation facilities is impeditive to development.

5. Child Labour System

Children are flower of a nation. They are future of a country. Therefore, development of a country depends upon the development of children. But in our country children belonging to below poverty line families do not get occasion of educational development. Many of them never see the face of school because they never get opportunity of enrolment in school. Instead of attending school, these children go to render their services as child labour. For them earning bread looks more important for survival than educating themselves for development. As child labour they are exploited, abused and deprived of human rights.

6. Bonded Labour System

Bonded labour system is a system in which a labourer enters into the agreement with an employer to render his services on loan till the loan plus interest is not paid back. In this period, he gets nominal wages. He forfeits his freedom to choose another employment and bargain his wages. Loan is the root cause behind the existence of the system. This system is beneficial for the employer and exploitative for the labour.

There is provision in this system to become free by paying back loan and interest. But in practice it is not possible to pay back loan and interest from nominal wages. Another point is that he incurs another loan before the expiry of the first loan. In this system a labourer is fully dependent upon the employer for his livelihood and survival. He cannot think of his own development or development of his family members. This labour system is impeditive to development.

7. Indebtedness

Indebtedness is a condition in which an individual takes loan from non-institutional agencies like the mahajan at exorbitant rate of interest for nonproductive purposes like marriage, feast, festival, treatment, consumption etc. Since the rate of interest is very high in comparison to banks and cooperative society, a man is forced to be exploited by the *mahajan*. The *mahajan* charges compound rate of interest. The man does not come out of indebtedness in his life time. After him, his loan liability is shared by his sons. Under the situation of indebtedness of a *mahajan*, a man cannot imagine about his development or development of his family.

8. Displacement and not Proper Rehabilitation

Displacement and not proper implementation of rehabilitation scheme also creates an barrier for development. It makes the poor displaced families more and more poor. It also pushes the displaced families towards extinction. Different evaluation reports submitted by different agencies dealing with establishment of industries, dam, thermal power stations, offices, railway, road, aerodrum etc. have revealed the problem of displacement and improper rehabilitation of uprooted families. These reports also mention extinction of a number of uprooted families. Some reports have highlighted uprooted families living in slum and leading life worst than animals. Some were forced to do beggary, theft, snatching and prostitution. Thus, displacement and improper rehabilitation have caused economic impediment to development of a number of uprooted families due to development projects.

IV. Political Impediment

Impediment to development is also caused by different political agencies and institutions. These agencies and institutions oppress the poor in different forms and make them more and more poor day by day. Being and becoming poor force them to remain away from the fruit of development. Let us discuss about these political agencies creating impediment to development one by one.

1. Land Lord

The landlord oppresses the landless in a number of ways. They employ the youth of landless families as bonded labour, attached labour, contract labour and casual labour. But they do not pay them minimum wages revised by the Government time to time. Besides they force labours to do *Begari* without any wages or food. They abuse them and torture them, too. The poor labourers do not resist because they have always fear not to get bread. The landlord compels them to render labour at back and call. Each landlord possesses his own labourers upon whom he has full right to exploit and oppress. The children of poor families are employed by landlord as herdboy on only food and clothes. They provide them bad quality of food and old cloth left by his family members. Thus, landlords as powerful and resourceful persons create impediment to development of the poor families.

2. The Mahajan

Mahajan is a local term used for those resourceful persons who provide consumption loan to poor families in rural and tribal areas. Many of them have government licence for doing moneylending. The poor families visit in the house of the *mahajan* to take loan for consumption purposes like food items, feast, clothes, festivals, treatment, marriage etc. The *mahajan* remains always ready to provide the poor families loan without any security because he knows each family members by name, face and tola or village. The *mahajan* charges high rate of interest on loan. After a year, he charges compound interest. Since banks and cooperative society do not provide loan for consumption purposes, hence poor families are bound to visit the house of *mahajan* for taking loan. The mahajan exploits the poor families in such a way that they never come out of his indeptedness. Once in indebtedness of *mahajan* remains always in his indebtedness. Indebted families are not in a position to think about development. The *mahajan* are powerful too. They have contact with thana, police and politicians. When an indebted family is unable to repay the loan plus interest, the *mahajan* confiscate land, cattle, ornament, tree, house, etc. If the concerned family has no property, children and family members are asked to render labour in his house or field on only both ends meal. Thus, the *mahajani pratha* is impeditive to development.

3. *Middle Man*

In the local haat, there exists middlemen. They purchase the things produced by the villagers. They also sell items of daily consumption and other articles to villagers. The middle men also oppress and exploit the poor villagers. They purchase the produce of villagers at very low price, but sell their own articles at very high price. In absence of money, the middle men also provide them articles on credit basis. But they charge interest very high. On being unable to repay, the middle men grab, the land, cattle, ornament, house, agricultural fields etc. They employ them as herd boy, domestic servant and bonded labour on nominal wages or food and clothes. As the loanee never comes out of clutches of the middle men, they can never think of their development.

4. *Choukidar*

Choukidar is a person who reports about crime situation in the villages to the Thanedar. Since he is associated with

Thanedar, he possesses some power amidst the poor villagers. Being powerful to some extent, the choukidar also oppresses the poor families who fear from thana-police and court. The choukidar takes away goat, hen, cock, egg etc. from the house of the poor. He threatens them to register their names as criminal in the thana. He also exploits the poor women physically. Thus, choukidar as a political agency creates impediment to development.

5. *Thana-Police*

Police stations have been established to control crime in an area so that people could live peacefully and develop peacefully, but thana police act adversely, too. They take bribe from the crime doers and provide them safety. In their names and places, poor and innocent persons are arrested and sent to jail as criminals. With the grace of police and politicians, crime doers wander freely on road, while innocent poor live in jail. The families of such innocent people become victims of oppression. They remain away from development. Thus, Thana-police create impediments to development.

6. *Traditional Caste/Tribe Panchayat*

It is true that traditional caste/tribe panchayat plays significant role in resolving problems of the caste or tribe concerned. But now-a-days, traditional caste/tribe panchayat has not remained impartial. It has started taking side of any one party. As a result, the real culprit does not get punishment. Innocent person is punished due to partialism. This creates impediment to development.

7. *Traditional Village Panchayat*

It is true that traditional village panchayat plays valuable role in bringing peace and harmony in the society. Thus, it plays valuable role in developing social harmony and peace by resolving village level issues through village consensus. But due to factionalism, traditional village panchayat has not remained partial. Now on many occasions decisions are not taken on consensus basis, but on the basis of majority of dominant group. This has taken away the very purpose of traditional village Panchayat in which every villager participated in consensual decisions. They were involved in decision making as well as in decision implementing as a member of community. Factionalism in traditional village Panchayat has created impediment to development of peace and harmony in the community.

8. Modern Gram Panchayat

Under Panchayati Raj Administration, Modern Gram Panchayt System has been established to provide administration and implement development programme at panchayat level. At Block level, there is Panchayat samiti and at district level, there is Zila Samiti. For the purpose of election of these political institutions, elections are held. The election of these institutions invites factionalism and feuds. As a result, new political equations are formed. These factions and feuds continue even after the election and declaration of results. Factionalism and feuds appear in Gram Sabha meeting, formulating and implementing rural development programme. The winner factions try to implement more and more development programmes in their villages and *tolas*. The runner factions are deprived of development programmes. Thus, elections of modern Gram panchayat has not only destroyed the development of social harmonial relations, but it has also created impediment to development of runner factions.

9. Election of MLA and MP

On the occasion of elections of MLA and MP, ugly scene of factionalism, casteism, communalism, regionalism and linguism, etc. appear. Incidences of caste riots, communal riots, regional riots and lingual riots also appear. These incidences destroy not only social peace and harmoney but also hamper national development. On the occasion of election booth capturing, booth looting and riots take place. They cause harm to lives, individual, property, cultural heritage and national property. Thus, on many occasional, many MLA and MP. Constituency elections create impediment to development.

10. Political Parties and Leaders

Different political parties and leaders also create barrier for development. With the help of their supporters, they organise bundh. The bundh supporters come in conflict with traders, transporters, servicedoers, daily wage earners and administration. They are also involved in loot, arson, breaking, fighting etc. They also cause harm to railway, bus stand, Government offices and national property. Schools and colleges are closed. Plying of rails, buses and trucks are stopped. Daily wage earners cannot earn to eat. Transanctions are stopped. Such damages created by different political parties and leaders are detrimental to development. In this way political parties and leaders create impediments to development.

11. Attrocities on Harijan

Harijans have been powerless and resourceless in Hindu social structure. After independence, our Central and State Governments have attempted to uplift their social, economic, educational and political status. These events are not acceptable and suitable for those Hindu castes who occupied top or middle position in caste hierarchy. As a consequence, Harijans have been subjected to attrocities in the form of house burning. thrashing floor burning, confiscation of property, cattle wealth, house, home stead land, looting, beating, killing, mass killing, rape, gangrape etc. Such behaviours are not congenial for the development of Harijans. Thus, attrocities have resulted in impediment to development among the Harijans.

12. Sexual Harassment of Women

In our society, women have been powerless and resourceless. They are subjected to sexual harassment in journey, marketing, fair, market places, railway stations, bus stands, work places, cinema hall, ration card shops etc. Being feared from sexual harassment a number of women do not like to travel, go to schools, colleges, market place, fairs, railway station, bus stand, work place alone. Sexual harassment of women is impeditive to their development.

13. Atrocities on Women

Being power less, women are subjected to atrocities in the form of infanticide, foeticide, bride-burning, poisoning, dowry death, drowning, dowry death etc. In this way altrocities on women are detrimental to their development. Atrocities are impemdiental to women development.

14. Not Participation in Democratic Process

Women have to observe a number of taboos. In case of forward castes, they are not allowed to go to cast their votes. They are not allowed to become a candidate of *mukhia*, Panchayat Samiti Zila Parishad and other elections. Thus, they are prohibited from participating in democratic process by the society. This practice is harmful for the development of women.

15. Court

Courts have been established for ensuring judgement to the people. But now-a-days, going court for judgement has become a costly affairs. Besides judgements are sold and purchased in the courts. Wealthy persons on the basis of wealth purchase judgement in their favour. The poor people being unable are to pay bride are unable to get judgement. The sale and purchase of judgement is detrimental to development process.

V. Religious Impediments

Religion is belief in superhuman or supernatural powers. These powers control and guide the behaviour of human beings. All events good-bad, right-wrong, joy-sorrow, advantage-disadvantage, fame-defame, life-death etc. are attributed to these superhuman powers. Development of individual, society and nation is also attributed to super human powers which are known by various names such as Gods, Goddesses, ancestors, spirits etc. According to our belief system economic, educational, health, political and social developments are gifts of deities. Without their blessings development of an individual, society and nation cannot be imagined. But contrary to these beliefs religion is impedimental to development as well. Impeditive role of religion to development can be discussed under following headings for the purpose of our study and knowledge:

1. *Superstitious*

Belief provides us force and we move forward. But along with beliefs, some superstitious are also prevalent in our society. Superstitious are detrimental to development. In some societies superstitions regarding human sacrifice are prevalent. It is believed that human sacrifice will bring rain, increase fertility of land and bring good harvest. There is superstition that when naked women plough the field, God Indra becomes pleased and gives good rainfall and good harvest. A superstition is prevalent in the rural areas that sprinkling of blood in the field increases fertility and yield. Again there is superstition that women, shudra and drum are meant for beating. They can be made refined by beating. Similarly a number of superstitions are prevalent related to journey, education, business, agriculture, animal husbandry, goatery etc. Prevalence of such superstitions creates impediment to development.

2. *Witchcraft*

A number of superstitions are prevalent regarding the witchcraft. Witchcraft is related to woman who is widow, childless or has only daughters. As she is widow or childless or possesses only daughters, she is frequently termed as witch.

She as a witch is looked down in the society. She is supposed to bring disease, illness and death in a family. She is abused and forced to walk naked in the village with head hairs shaved. She is abused as witch only to grab her husband's property. Thus label witch is detrimental to development of a woman. It creates impediment to her various kinds of development.

3. *Sorcery*

In society superstitions are prevalent regarding some persons as evil eyed, evil-mouthed and evil-faced, evil-toothed, evil-tongued and evil footed. There is superstition that such persons possessed evil spirits in their mouth, eye, face, tongue, tooth, legs etc. with the possession of evil spirits, they perform black magic to destroy a person or his/her entire family members. A sorcerer may be a male or a female. Sorcerers are looked down in the society. They bring quarrel between families. They destroy peace and harmony of the society. They create impediment to development in society.

4. *Shaman*

Shaman is a person who possesses a powerful benevolent spirit. With the help of that powerful spirit, the shaman neutralises the effect of witch and sorcerers. In local terms he is known as *Ojha-Guni*. He or she is an expert of driving out an evil spirit. But a shaman brings conflict and quarrel in the society by declaring a woman as witch or a man as sorcerer. Bloodshed and murder also take place. In this way shamanism is impedimental to development of society.

5. *Religious Rituals*

Religious rituals are performed with the help of specialists and materials of worship. This requires money as well as time. Sometimes, specialists force a family to spend more and more on ritual performance for various kinds of benefit in the family and society. On many occasions a person has to take loan from the mahajan for the performance of rituals. He becomes so indebted that he could not come out of indebtedness. In order to repay the loan, he has to sell or mortgage his land, tree, cattle, ornament, etc. In this way performance of religious ritual on taking loan becomes impeditive to development.

6. Extravagency on Marriage Ritual

In Hindu society, marriage is a religious sacrament. A number of duties and ancestral spirits are worshipped on the occasion of marriage to seek their blessings for peace and prosperity in married life. On this occasion a grand feast is given to friends and relatives. Exchange of gifts and presents also take place on this occasion. Extravagency on marriage rituals brings a person under indebtedness of the *mahajan*. In order to pay loan and interest, he has to sell or mortgage his ancestral or earned property—house, land, tree, cattle, ornament etc. If he is unable to repay loan from these properties, then he has to repay by working as bonded labour. In this way, extravagency on marriage ritual is impeditive to development.

7. Heavy Expenditure on Death Rituals

There is tradition to spend heavy on death feast and rituals. For this purpose, majority of people take loan from the *mahajan* on exorbitant rate of interest. The *mahajan* charges compound rate of interest. In order to pay back the loan and interest, one has to sell or mortgage his house, land, tree, ornament, cattle etc. If not, he has to render labour on nominal wages along with his wife and children. Thus, religious traditions make a person indebted and indebtedness brings impediment to development for the concerned family.

8. Pilgrimage

In our tradition, the pilgrimage has religious value. Pilgrimage needs money and time for journey. It is a costly affair. But it is so valuable religiously that every aged person of the family want to make pilgrimage in his/her life time. For pilgrimage people take heavy loan from the *mahajan* on exorbitant rate of interest. If unable to repay, they have to sell or mortgage their house; land, tree, cattle, ornaments etc. in the hand of *mahajan*. Thus, pilgrimage makes indebted and indebtedness creates impediment to development of various kinds.

9. Festivals

We have a rich tradition of celebrating festivals with pomp and show. For this purpose we have to do marketing of new clothes, items of worship, items of decorations, food items and drink items. Majority of people take loan to celebrate festivals. Thus, celebration of festivals makes people indebted to the *mahajan*. And on becoming indebted and unable to repay, he/she has to sell or mortgage his property which retards his development to a great extend.

10. Donation of Goods and Services

In the name of religion donation of goods and services are made. Ricen people donate money, rice, ghee, *dhup*, loafsugar, sugar etc. The poor people donate their service. There is no harm in donating goods and services in the name of own religion. The rich and the poor both can do it according to their will. The harmful is taking donation and services forcefully. For temple erection, doing *yagya* and digging out a religious pond, even poor people are forced to pay money and to render labour. Paying money and services under compulsion is not good from the point of view of development. It is rather impeditive to various kinds of development.

11. Capturing Land in the Name of Religious Centres

The followers of different sects and communities try to capture government land or to erect symbols of their religious deities. Thus, we find the existence of temples, mosques, Gurudwaras and churches built on government land. This creates difficulties in vacating the captured land. Because the followers organise themselves to launch a movement against the administration. As a consequences, road development, communication development and other kinds of development become impeditive.

12. *Communal Riots*

Each religion teaches about Gods and Goddesses. Each religion teaches us about peace and prosperity in the society for human development. But in spite of all such beliefs prevalent in society, we also witness the ugly scene of communal riots. In communal riots, people forget about peace, prosperity, harmony, service to man etc. They become loyal to their own religion. They try to cause harm of life and property to other communal group. At the time of communal riots, several persons are killed, their houses are burnt. Their properties are looted. Bloodshed, arson, loot, murder, attack with lethal weapons etc. become common in communal riots. On these occasions, people forget their national and human loyalty. They become loyal to their own community. Thus, communal riot is impeditive to development of individual, family, society and nation.

13. *Religious Value of Population*

Our Government has Ministry of Family Welfare. This ministry emphasises upon the small family size. It teaches and trains the couple for adopting birth control measures. But it has not been able to formulate a national population policy. Because, it is being opposed by different communities who believe in more and more increase in population. In their belief offspring is boon of Gods and Goddesses. According to them God is giver of mouth; hands and food to eat, grow and develop. On this ground, they oppose small size family and national population policy. But it is true that a big family has to struggle more for the development of its members. It is impedimental to development of children, economy and health in the family.

14. Idleness

Some people sit like an idle in the name of religion. They think that God will send him food to eat because he has given him birth. According to their belief birth giver is also food giver. Such beliefs and behaviours prevalent in the mind of people is really impeditive to development.

VI. Psychological Impediment

Psychology of individual or group plays significant role in development. If an individual is psychologically fit, only then, he can participate in development process. Bur if he is Psychologically unfit he cannot participate in development. For development, peace and presence of mind is must. A man who thinks positively and is optimistic participates properly in different areas of development. Psychological factors contributing towards the impediment of development can be discussed under the following headings:

1. Fear

Fear plays negative role in development. Due to fear, an individual does not take courage to take loan from banks or co-operative societies for education, small business, big-business, house construction, marriage of daughters etc. The fear is if he will not repay the loan what will happen to his family and paternal or earned property. Such type of psychology prevails in the mind of poor individuals. As a result, they remain away from educational, economic and other kinds of development. Some individuals have fear from success. That is why they do not appear in examination and undertake risk. Due to fear, they do not take risk of any kind. In this way, fear is impeditive to development.

2. Anger

Anger is a psychological factor which plays negative role in development. An individual must possess control over his/her anger. Because it prevents an individual from thinking in wrong direction. It brings quarrel and anoids other members of the family and community. When an individual becomes angry, he looses his intelligence and is bound to do such activities which are harmful for his own health as well as the community. Thus anger is impeditive to development.

3. Hatred

Hatred also creates impediment to development. It is considered not good for individual and social development. It destroys the own health of individual as well as it creates disharmony in the society. When an individual has hatred for someone else, he/she destroys his/her own energy for no gain. He/she does not like to see, talk and face the persons for whom he/she possesses hatred. He/she always think something wrong about those persons. This kills the time as well as energy of an individual. He/she lags behind in different areas of development. Thus, hatred is impedimental to development.

4. Jealousy

Jealousy is also not taken good for the development of own health as well as the health of community. Development also brings jealousy for the others. Many people are not happy because their neighbours are developed, peaceful and happy. Jealous people try to create harm to other members in a number of ways. They burn their own blood to cause harm to others. They kill their own time and energy for that purpose. As a result, they remain away from own development. In this way jealousy is impeditive to development.

5. Enemity

Enemity is a psychological factor which is not taken good for social peace, harmony and development. It is also not good for economic development. It brings quarrel between families. People take help of thana-police and court in case of enemity.

They have to spend huge money on cases in courts. They have to meet expenses on lawyers and journey. When enemity is developed, every kind of relationship is broken. Exchange of talk and relationship is suspended. Both groups always think to harm each other. Bloodshed and murder also take place in case of enemity. Thus, enemity is impeditive to development.

6. Frustration

Frustration is a psychological state of an individual which makes him/her passimist. Frustration develops abnormality in an individual. He/she starts behaving like mad or criminal. Frustration leads to antisocial activities like theft, dacoity, arson, rape, murder and even suicide. Thus, frustration creates impediment to development of various kinds.

7. Depression

Depression is a psychological state of an individual. The life of individual is full of events. Some events are pleasant and some are not pleasant. Unpleasant events create depression in an individual. Death in family, separation or failure in love, in attempt or make an individual depressed. In depression, an individual becomes either aggressive or submissive. His behaviour does not remain normal. He began to behave like a mad and makes several attempt to end his own life. In this way, depression is impedimentive to development of an individual and his society.

8. Prejudice

Prejudice is also a psychological state which creates the feeling of superiority and inferiority. Prejudices create impedimentive situation for development of society. In our Indian society various kind of prejudices are found. For example, racial prejudice, caste prejudice, religious prejudice, political prejudice. These prejudices create the situation of conflict and tension in the society. They hamper the development process.

9. Tension

Tension is also a psychological state of mind which hampers the development of an individual and his society. Tension may be of various kind such as familial tension, caste tension, communal tension, regional tension, lingual tension, etc. In the situation of tension, an individual cannot think properly. He cannot take right decision. His behaviour becomes abnormal. He expresses anguish over someone and when cool down he seeks excuse. Tension also results in antisocial activities like rape, arson, brickbating, murder, suicide etc. These abnormal and antisocial activities are not congenial for development of individual and society.

10. Revenge

Revenge is also a psychological state of an individual which is not taken good in the society. The revengeful attitude makes a man mad. He forgets every thing when he passes through the stress of revenge. He forgets his intelligence and humanity. He spends his time and energy in making plan to take revenge. Revenge is taken beating by beating, bloodshed by bloodshed, abuse by abuse, scolding by scolding, rape by rape, theft by theft, dacoity by dacoity, murder by murder etc. In revenge, people forget about pardon. This is harmful for individual and his society. Thus, revenge is impedimental to development.

VII. Natural Impediment

Impediment to development is also caused by nature. There are some natural events which cause damage to lives and properties and are detrimental to development. Natural events are not under the control of man. They are caused by nature. For our purpose of study and description we can divide the natural impediments to development under the following headings:

1. Flood

Flood is a natural event which occurs due to continuous and excessive rain. In case of flood, the water of rain is not flown by rivers. As a result, rain water spreads in villages situated on the bank of rivers. The flood water fells house, washes villages, kills people and cattle and causes damage to agricultural crops. Flood water also washes away bridges, roads, railways, electric poles, telephone lines etc. Thus, the flood caused great damage to development efforts already made. It needs more money to rehabilitate the flood victims. In this way, flood creates impediment to development.

2. Femine

Femine is a natural event. It occurs when there is no rainfall in rainy season. The farmers are unable to do agriculture on their lands in absence of rainfall. It brings the question of existence before farmers, agricultural labourers and cattle they domesticate. The situation of draught also brings problem of drinking water because the water level in the situation of femine goes down to such an extent that hand pumps and wells become non-functional. In order to save life people migrate from one place to another. In such a situation, they cannot bargain their labour at market value. Rather they are bound to work even on food or nominal wages. In draught people forget about education and standard of living. They want to save themselves by hook or crook. They adopt immoral means too. They are involved in theft, docoity, loot, rape, prostitution, etc. In this way, femine is impedimental to development.

3. Sea Storm

Sea storm is also a natural event. It causes sea water to enter in the villages situated nearby the sea-shore. Villages and towns situated on sea-shore are drawned and washed away by the sea storm. Sea storm blows away and washes away lives, cattle, trees, crops, houses, building, road, railway track, bridges, telephone lines etc. It makes drinking water dirty. It creates ugly scene of destruction on the earth. Damage caused by this natural event cannot be compensated by rehabilitational efforts of the Governmental and non-Governmental organisations. In this way sea-storm is impeditive to development.

4. *Thunder*

Thunder is a natural event which causes damage to lives of people, cattle and tree. The incidence of thunder takes place during rainy season. It can fell on any place and on any people. When it falls on an individual, he dies on the spot. Thus, the incidence of thunder is impeditive to development.

5. *Bursting of Clouds*

Bursting of cloud is also a natural event which causes damage to lives and property. Thus, this natural event creates impediment to development.

6. Earthquakes

Earthquake is such a natural event which causes damage to several lives and huge property in few seconds. Thousands and thousands of people are killed and huge amount of properties are destroyed. Thus, earthquake creates ugly scene of destruction on the earth. So, it is impeditive to development.

7. *Epidemics*

Epidemics are also natural event. They occur soon after flood, famine, sea storm, earthquakes etc. They are pox, cholera, plague, etc. and many viral diseases. If they are not tackled properly, they cause damage to several lives. In this way, epidemic are impedimental to development.

8. Administrative Impediment

Our country is a democratic country. The form of Government in our country is democracy. In our country there are bureaucrats and officials to run the Government. Our Government makes policies of development, formulates schemes of development, implements the scheme of development, monitors the development schemes and evaluate the development schemes with the help of Governmental agencies. There is no doubt that these Governmental agencies are involved in development, but it is also true that these agencies also create impediment to development. They do not implement programmes timely. They do not release fund easily. They impose development programmes upon the people without taking in to account their cultural needs. They do not follow-up a number of programmes after specified period of implementation. Thus, beneficiaries are left in doldrums. In this way, development has to face administrative impediment, too.

IX. Corruptional Impediment

The simple meaning of corruption is bad conduct. It is considered bad conduct because it involves those behaviours which are morally wrong. It is an act of characterlessness, moral degradation and unlawful practices. Corruption is related to earning money and positions by hook or crook without caring for duties and responsibilities towards people, society and nation. In our country corruption is all pervasive and omnipresent: India is large democracy of the world. It is second populous country of the world. India had cherished honesty, morality and high values from time immemorial. But with passage of time and fast socio-economic changes corruption has affected all walks of life. It is also true that a number of public servants, politicians, Government servants, bureaucrates, industrialists, traders etc. have been sent to jail on corruption charges. But these punitive steps, too, have not been able to put a check on this social evil. People are still compromising with principles. It has taken away our social virtues and has made us selfish and inhuman. It has almost become a way of life. The practice of nepotism, favouritism, dowry, dishonesty, immorality and foeticide are examples of social corruption. Payment or illeagal gratification for getting workdone in offices is the most common form of economic corruption. Making money from misuse of office and power to win election, utilising criminals to win elections, looting polling booths, giving tickets to criminals, giving bribe to legislators to get required number for making or unmaking of Government etc., are political form of corruption. Taking bribe and commission for getting fund sanctioned and released, or looting public money in development works, are examples of bureaucratic and administrative corruption. Admission on donation, free degree, diploma and certificate, mass copying, leak of question papers, bribing examiners, changing copies and results, supplying answer materials from outside and all kinds of cheating related to competitive examinations are examples of educational corruption. Corrupt practices prevalent in development programmes are examples of developmental corruption. Smuggling, black marketing, hoarding, profiteering, food adulteration, production of duplicate items and spurious medicines are examples of commercial corruption. Corruption prevalent in judiciary, medical institutions and service commission reveal professional form of corruption. Raising funds in the name of temples, mosques, churches and Gurudwaras, religious discources, performing yagya and celebrating festivals but spending otherwise reveal religious form of corruption. On the basis of above accounts it is clear that corruption is impedimental to development.

X. Educational Impediment

Education is an important means of development. Educated people are able to get Governmental or non-Governmental jobs. They become professionals too and earned good money. Uneducated people are not in a position to get white collar jobs. They do not become professionals. They earn only as casual labourers. Education is also necessary for better understanding of a problem. It is essential for national and international level communication, business, jobs and professions. Education provides us varieties of opportunities for jobs, occupations and professions. But unfortunately we have not been able to make all people in the ages 7 years and above as literates. Literacy rate in our country is nearly 65 per hundred. For males it is 75.26 per cent and for females it is 53.76. This is national figure. This figure differs communitywise and statewise. Educational impediment to development can be discussed under the following headings for the purpose of study and description:

1. Non-enrolment

It is a matter of great concern that even after 59 years of achieving our independence our people residing in remote areas and rural areas are facing problem in enrolment of children. Even today there are many families whose children have not seen the face of school. For them a school is nothing more than a building to be seen from outside. Enrolment is basic condition for making an individual literate. Non-enrolled children in school are bound to live illiterate throughout their life.

2. Non-attendance

It has been observed that a number of or a good number of enrolled children start non-attending their classes. Non-attendance of classes shows lack of interest in education. It seems that children were enrolled forcefully or enrolment was imposed upon them. Non-attendance provides a ground for drop-outs.

3. Drop-outs

As referred to earlier the beginning phase of drop-out starts off and on non-attendance. Finally, they start remaining absence regularly. Their regular non-attendance results in drop-outs. Till the phase of drop-out some are lucky to become just literate.

Thus, non-enrolment, non-attendance and drop-out are impeditive to development of education.

4. *Not follow-up of Educational Programmes*

A number of children of poor families leave the school because they sometimes receive reading materials, dresses, midday meal, scholarship and even accommodation in school hostel. But after some period, they do not get these things due to shortage of funds under educational development programmes. Such incidence creates adverse effect upon the educational development.

5. *Truancy*

Truancy is detrimental to educational development of children. Truants are those children who leave the school without any information. They remain absent from school without permission of parents and school authority. They neither attend school nor return home in school hours. They fall in bad company. They come in contact with criminals. For them, truancy creates ground for becoming criminals. Thus, truancy is impeditive to development.

6. *Juvenile Delinquency*

Truants come in contact with bad persons. They gradually leave school for ever. They become juvenile delinquents. As juvenile delinquents they are caught and beaten by the people, police and other persons. They are sent to jail. Their personality is disorganised. Thus, juvenile delinquency creates impediment to development.

7. *Vagarancy*

In truancy, the school going children fall in bad company. They do not attend school. In school hour, they wander in markets, on road, station, bus stand etc. and make vulgar comments on girls and women. They are seen in *puja pandal*, political processions, religious processions and crowdy places. As vagarants, they are not only problem of family concerned, rather they create problem for the whole society. Sometimes, they are beaten mercilessly. They are also sent to jail. In this way vagarancy is impeditive to development.

XI. Health Impediment

Health is most valuable wealth. Without health a man is worst tham an animal. A healthy person participates in development process of society and nation. He makes contribution not only towards his own development, but for the development of his society and nation, too. Unhealthy persons become liability for family, society and nation. But there are many factors which are impeditive to health development. We can discuss health impediment to development under following headings for the purpose of our study.

1. Malnutrition

Nutrition is basic condition of health. Without nutrition one cannot participate in his own development, familial development, societal development and national development. Malnutrition affects the health adversely. It makes an individual weak. He as a malnutrient does not participate in his own development. He remains aloof from familial, societal and national development. A malnutrient woman delivers a dead child or an unhealthy child. The question of survival appears before unhealthy baby. Her life is also threatened during pregnancy. Thus, malnutrition affects the entire development process and creates impediment to development.

2. Population Explosion

Population explosion is impedimental to not only health development, but development as a whole. Big family has to suffer from malnutrition. Big family is not in a position to spend more on education of children. Big families do not send their children in good school for education. They do not provide good space to live and sleep. Big families have to struggle hard to maintain their family members. They do not spend much on health, illness and treatment. They remain always under debt, so they do not participate properly in development process. Thus, population explosion creates impediment to development.

3. *Unhygienic House*

Families belonging to below poverty line generally live in unhygienic house. Their houses do not have windows, kitchen, latrine, urinals, drainage, bathrooms etc. They generally live in single room with vèrandah where they have a store, kitchen, cowshed, etc. In one corner they have a pit for washing utensils. This pit is also used as urinals. They ease on street outside the house. Unhygienic house affects their health. They become victims of a number of diseases. Thus, unhygienic house creates impediment to development.

4. *Unhygienic Workplace*

Poor labourers have to work at unhygienic places in factories, mines, industries etc. At those places they have always risk to their health and life. Thus, unhygienic work place is impeditive to development.

5. *Lack of Drinking Water*

In our country many remote hilly, forested, deserted and rural areas lack drinking water facilities. On those places, women children and even men spent their most time in attempt to get drinking water from other places. They have to travel one to five kms for bringing drinking water. Lack of drinking water make them reluctant towards development efforts and processes of the Government. They do not get time for education and other kind of development. They do not take sufficient drinking water which creates health problems for them. Thus, lack of drinking water is impedimental to development.

6. Lack of Health Education

Lack of health education is impedimental to development. Health education helps us in building good health habit. It also makes us aware of several health problems and their remedial measures. In absence of proper health education, we follow a number of bad habits which are detrimental to our health, wealth and prosperity. Thus, lack of health education creates impediment to development.

XII. Terroristic Impediment

Terrorists are those persons who want to terrorise the people by their heinous crime. They make bomb blast at crowdy places and trains. They hijack aeroplane. They destroy cultural and national heritage. They attack on professional centres, market centres, big and multistorey building, Government Offices, share market centres, etc. They gundown villagers, tourists, pilgrims etc. Terrorists create great damage to a nation by their inhuman and heinous activities. Our country is facing terrorists attack since decades. They have caused a great damage to lives and property of people as well as wealth of nations. Thus, terrorists create impediment to development.

XIII. Ecological Impediment

Development has to face ecological impediment, too. Ecology of different areas is not similar. Ecology of plain area, desert area, costal area, hilly area, forested area, cold area etc., differ profoundly. Undoubtedly, there has been relation between ecology and technological development. In each particular ecology, a particular technology is suitable. As a result, a particular type of social organisation emerges there. Mountain, sea, desert, lake, rivers etc. create impediment to uniform development everywhere. They bring variation in regional development or area development.

XIV. Communicational Impediment

Communication is vehicle of development. It makes us conscious towards various kind of development. It also makes us aware of ongoing programmes of development. Government makes people aware of ongoing programmes through different types of media such as folkmedia, mass media, print media and electronic media. But it is unfortunate that people are suffering from communicational impediment to development. Government officials connected to the villages are school teacher, VLW health servants and panchayat sewaks. They do not communicate the people properly about the ongoing programmes of development. They neither communicate timely nor convince them about the result and benefit of development programme. When people become aware of the programme, the period of programme ends. People face communicational problems, too regarding various programmes. They are not informed in local languages and through folkmedia like sloagan shouting, procession, song, dance, music, drama, conversation, etc. Thus, Lack of proper communication about development programmes is impedimental to development.

XV. Naxal Impediment

In our country certain states and area have become naxalite infected. Naxalites are running parallel Government in those areas. They commit organised crime like group killing or massacre. They make attack on police pickets, police thana, railway station, railway bridge, railway track, with bombs, landmines, AK47 and other advanced weapons. They demand levy from the villagers. They also demand family members to work for them. They kill landlords, police informers and others who come in their way. They make loot of police weapons and attack on jails to release their fellows. They also have female party workers. They take shelter in villages as forceful guests on barrel point. They demand commission from BDOs, SPs, DMs, DCs ministers, MPs and MLAs for smooth implementation of development programmes. Without taking commission, they do not allow to implement development programme in the area. They held property in disguise names. They have huge money in bank accounts in different states. They are also owners of multistorey flats in state capitals, and important cities. They stop train, buses, transportation of goods, schools, plying of railways, functioning of telephone exchange, etc. Schools teachers and doctors do not attend their duties in the villages from their fear. In this way, naxalites create impediment to development.

7

Governmental Agencies of Development

Today we have state as a most evolved form of society. Each state has boundary demarcation, population living in the boundary, Government to protect the boundary and the people and sovereignty. State provides citizenship and fundamental right to its citizens. To bring all people in the main stream of development is the prime duty of Government of a state. Our country India is a democratic and secular state. It is the prime duty of our Government to bring all people in the main stream of development beyond caste, race, creed, region, religion, language, gender etc. After independence our Government is functioning in this direction. As a result we have a number of Governmental agencies involved in the development of people society and nation. Governmental agencies of development can be discussed under the following headings:

1. Gram Panchayat

Gram Panchayat is basic Governmental Agency involved in the development of villages and their people. The head of the Gram panchayat is *mukhia*. The *mukhia* is elected through ballot papers after each five years in democratic manner. Each panchayat has a boundary and voters of 18 years and above. A panchayat has several villages under its jurisdiction. Now-a-days, panchayat have been made responsible for formulation, implementation, monitoring and evaluation of rural development programmes. Panchayat *mukhia* posts have also been reserved for women. Main parts of a Gram Panchayat are *Gram Sabha* (Village Assembly), *karyakarini samiti* (Executive Council), *Gram kachehari* (village court), *Gram Sevak* and *Gram Raksha Dal* or village police. The main source of income of Gram Panchayat is Government fund. The main function of Gram Panchayat is alround development of the village.

2. Janasevak or Gram Sevak

Each Panchayat has a Panchayat Office. The custodian of this office is called Gram Sevak or Jana Sevak. Jana Sevak is a Government Servant. He/she receives monthly salary from the Government. Jana Sevak is a connecting link between the state Government and the Panchayat. Gram Sevak keeps records of the Executive Committee, Gram Sabha, Gram Kachehari and Gram Police.

3. Panchayat Samiti (Panchayat Council)

It is also known as Prakhand Samiti. In each Prakhand (Block) there is Panchayat Samiti or Prakhand Samiti. It includes all Mukhias of Panchayats coming under the jurisdiction of a Block. Members of Panchayat Samiti elect one person as Pramukh and one person as deputy Pramukh. Block Development Officer is secretary of Panchayat Samiti. MLA, MLC, MP and member of Rajya Sabha are associate members of Panchayat Samiti. The Panchayat Samiti has at least two women members. Two experts of rural development can also be included in the Panchayat Samiti. One member of managing committee of cooperative bank, three elected members by secretaries of cooperative societies and deputy chairman of municipality and scheduled area committee are also members of Panchayat Samiti. The main function of Panchayat Samiti is to establish coordination with Gram Panchayat, cooperative societies, voluntary organisations and people for community development.

4. Zila Parishad

According to Panchayat Raj Adhiniyam, in each district there is a Zila Parishad. A chairman and deputy chairman is elected from among the members of the Zila Parishad. DM or DC is the Secretary of Zila Parishad. It establishes coordination between Gram Panchayats and Panchayat Samiti in development works. It also establishes coordination in plans formulated by different Panchayat Samiti of a district. It also monitors different development programmes going on in each block of the district. It makes its advice to State Government regarding the development of Panchayats, Prakhands and district. It uses its right provided by the Government. It verifies the budget of the Panchayat Samitis and makes distribution of funds provided by the Government among Panchayat Samitis. It formulates plan for the development of the district. It also makes collection of essential data for development work of blocks and district. It also helps in the establishment of professional and industrial schools.

5. Schools

Our Government has established Primary Schools, Middle Schools, Secondary Schools and plus-2 Schools to impart education to the children. Cent-per-cent expenditures on infrastructures, teachers and staff are met by our Government. Some of these schools are residential. Although co-education is prevalent in these schools, but some schools are specially Government Girls Schools. The children do not have to face any fees in these schools. STs, SCs and OBCs children get reading material, dresses and midday meals. General students at primary level also get these facilities. Some children get scholarships, too. School buildings are erected by the Government. The salary payments of teachers and staff are made by the Government. The Government has made attempt to establish primary level schools in all villages or in the radius of one km. A middle school in radius of three km and a Secondary Schools in radius of five km and plus-2 level Schools in each Block. The teachers of these schools not only teach the student, but they also help the villagers in solving their various problems. They extend help in census survey and elections of MLA, MP, Mukhia, etc. The schools are very important

Governmental Agencies related to educational development of children.

6. Colleges

Our Government has also established colleges to impart college level education to students. Expenditures on infrastructures, teachers, staff are met by the Government. Constituent colleges are cent-percent funded by the Government. Affiliated colleges receive financial support from the Government. They are also made constituent unit after fulfilling certain criteria laid down by the Government. Although co-education exists in every colleges, but some women's colleges have also been established. Some minority colleges also exist. Colleges play significant role in development of college level education in different subjects. Some colleges are only intermediate colleges, while some other colleges are degree colleges. Fees paid by students are nominal.

7. Technical Colleges and Institutes

Our Government has also established technical colleges and institutes to impart technical education to students. Government has established Engineering colleges, Indian Institutes of Technology, Regional Institutes of Technology, Medical Colleges, Law Colleges, Indian Schools of Mines, Indian Institute of Management etc. The expenditure on infrastructure, teacher and staff are met cent-per-cent by the Government. Students have to pay some amount as tuition fees, accommodation fee, development fee etc. but they are nominal in comparison to private institutes and colleges.

8. Universities

For the purpose of higher education and post graduate teaching our Government has started universities. The expenditure on infrastructure, teachers and staff are met cent-per-cent by the Government. Students have to pay some fees to get admission in universities for studying various subjects as they like to study.

9. Health Sub-centres

Our Government has established health sub-centres jointing two or three panchayats together. In this primary health sub-centre, there are two posts. One post is for a female health worker and the other post is for male health worker. Block medical officer remain in incharge of these health sub-centres. In newly created health sub-centre, there is only one post and that too for female health worker. Cent-per-cent expenditure on infrastructure and staff are met by the Government. These health sub-centres play significant role in development of health of the rural people. They also make people aware of health problems and their remedies through treatment and control. They also impart health education and training for family welfare programme.

10. Primary Health Centres

In each Block, there is primary health centre. In each primary health centre, there are four posts of medical officer, besides nurses and other staff. Expenditures on infrastructures doctors, staff, medicine, material etc. are met by the Government. These primary health centres play significant role in imparting family welfare programmes, health education, sex education, etc. They play significant role during epidemic control.

11. Extra and Subsidiary Health Centre

Extra and subsidiary health centres have been established at 20,000 to 30,000 population. There are two posts of medical officer in each extra and subsidiary health centres. Expenditure on infrastructure, medical officers and staff are met by the Government. People get free treatment here. They are also made aware of family welfare programme and other programmes related to health of children, mothers and community health.

12. Referral Hospitals

Our Government has established referral hospitals, too. In each referral hospital, there are four doctors. One is child specialist, second one is Gynaecologist and female disease specialist, third one is surgeon and fourth one is expert in medicine or general physician. Here, too, treatment of the people is met free of cost.

13. Sadar Hospitals

In each district, there is a sadar hospital with doctors as expert of different branches. This hospital is equipped with more medical facilities. People get here treatment free of cost. Expenditures on infrastructure, doctors and staff are met by the Government.

14. Government Medical Colleges and Hospitals

Government has established medical colleges and hospitals for teaching and training of medical students in different branches. Here, there are outdoor and indoor treatment facilities either free of cost or on nominal charges. Patient admitted get medicine and food, too, besides treatment.

15. All India Institute of Medical Sciences

In the national capital, Delhi, there is AIIMS for teaching and training of medical students as well as treatment of outdoor and indoor patients coming from different parts of the country.

16. Municipality

In small towns, municipalities have been established. It is established only in those towns which fulfil the following criteria:

(i) The town must possess at least a population of 20,000.
(ii) The density of population should be 1000 per sq.mile
(iii) 75 per cent people should possess non-agricultural occupation.

The structure of the municipality includes the following:

(i) Municipality Council

In each municipality council, there are members between 10 to 40 who are designated as ward commissioner. The number of council members depends upon the population of the town. Ward commissioners are elected by the people. Previously, one-fifth members are nominated by the Government. Now-a-days all ward commissioners are elected directly

(ii) Chairman and Vice-Chairman

Previously,all elected and nominated members of a municipality council, elect one chairman and one vice chairman from the members of the council. These posts are without salaries. Chairman is the leader of members of the council. He plays significant role in preparing budget of the municipality and getting it passed through the Government. He monitors the daily work of the municipality. Municipality has some permanent staff paid by the Government. Now-a-days, chairman and vice-chairman are elected directly by the people.

(iii) Committees

The municipality council establishes a number of committees to look after the development work of the town. These committees are eduction committee, finance committee, health and sanitation committee, public construction work committee etc. The chairman is president of all committees.

Functions of Municipality: For the purpose of study, the functions of the Municipality may be divided into two parts—(1) Essential functions and (2) Optional function.

Essential functions. The essential functions of municipality are as follows:

1. Arrangement of sanitation for town.
2. Construction and repair of roads in town.
3. Arrangement of drainage in town.

4. Arrangement of street light.
5. Arrangement of drinking water.
6. Arrangement of public market.
7. Arrangement of primary school.
8. Arrangement of hospitals.
9. Arrangement of cremation places and burials.
10. Sanitation of public places like field, parks.
11. Vaccination to cheek epidic.
12. Arrangement of meat shops, fruit shops and essential commodity shops.
13. Maintenance of death and birth records.
14. Arrangement for controlling fire.

***Optional works*:** The optional works of municipality depend upon the economic resources of the municipality. For example construction of new roads, drains, street, houses, etc. It can convert a dirty area into a livable place. It can make arrangement of land for construction of new houses. It can construct shop for distribution among the people. It can make arrangement of milk, meat, fruit, etc. It can make arrangement of parks, garden, museum, pond, library, rest house etc.

***Sources of Income*:** Sources of Income of municipality are customs, toll, tax, Government grant and Loan. The municipality has right to collect house tax, road tax, water tax, latrine tax, light tax, occupation tax etc.

17. Urban Area and Notified Area

The towns which have population less than 20,000 to 10,000 have Urban Area Committee and Notified Area Committee for the purpose of development of those towns. It has five to seven members. The committee functions like municipality.

18. Municipal Corporation

For big cities and metro polis, there are municipal corporation for development and administration. Municipal corporation has following parts:

(i) Corporation Council

(ii) Standing Committee
(iii) Chief Executive Officer

Corporation Council: The members of corporation council are termed as councillor. The number of councillor dependeds upon the population of the cities or metropolis. They are elected by the people. Some members are nominated by the State Government. Municipal corporation functions under the control of State Government. This council formulates policies of municipal corporation. Thus, it is policy making institution. The general meeting of the municipal council takes place once in a month. The special meeting of the council can take place when ever is required. The council has a mayor and a deputy mayor. Previously, mayor and deputy mayor were elected by the members of the council. But now-a-days, they are elected by the people.These posts are honorary. Mayor conducts the meeting of the municipal corporation. In his absence, deputy mayor conducts the meeting. The mayor is the first citizen of the city.

Standing Committee. This is second important part of municipal corporation. It conducts works related to the governance of the municipal corporation. The chief executive officer of the municipal corporation has to seek approval of many matters from standing committee. This committee has 15 members including mayor and deputy mayor. The meeting of this committee takes place once in a month essentially. Besides the standing committee, there are other committees, too, for extending help and cooperation in the functioning of the municipal corporation.

Chief Executive Officer: The municipal corporation has a chief executive officer. This officer is appointed by the State Government on the advice of the mayor. This chief executive officer belongs to IAS category. Chief executive officer is the chief administrator of the municipal corporation. He is responsible for implementing all acts and statutes of the municipal corporation. He brings wills and decisions of the municipal council and the standing committee into action. He can take part in the meeting, but he cannot participate in voting.

Functions of Municipal Corporations: The municipal corporation performs the following functions:

(i) Construction of drains and public latrines
(ii) Arrangement of sanitations
(iii) Arrangement of drinking water
(iv) Arrangement of street light
(v) Construction of cremation grounds and burials
(vi) Maintaining record of birth and death
(vii) To clean dirty places
(viii) Arrangement of Vaccination to prevent the spread of epidemics
(ix) Arrangement to prevent the spread of dangerous and contagious diseases
(x) Construction of public medical stores
(xi) Establishment of hospital, delivery homes and welfare centres
(xii) Construction of roads, bridges and drains
(xiii) Arrangement of primary schools library and reading rooms
(xiv) Arrangement to prevent the spread of cattle diseases
(xv) Arrangement to prevent the spread of fire
(xvi) Arrangement of public parks and gymnasium
(xvii) Construction and repair of banks of ponds
(xviii) Arrangement of transportation
(xix) Arrangement of exhibition and fair
(xx) Arrangement of markets and meat shops.

Besides these functions, municipal corporation can perform any work related to the interest of the people. It transacts all work from cradle to grave.

Sources of Income: Sources of income of municipal corporation are as follows:

1. Tax : The municipal corporation has right to collect following kind of taxes:

(i) Holding tax on house
(ii) Water tax, light tax, latrine tax and drain tax
(iii) Tax on rickshaws, bicycles and other kind of transport.

(iv) Bullock cart tax and tax on domestication of dogs
(v) Advertisement tax
(vi) Tax on sale and purchase of goods
(vii) Tax on markets and meat shops
(viii) Boat and steamer tax

2. *Grant-in-aid*: The Government provides Grant-in-aid to municipal corporation time to time.

3. *Fees and fines*: The municipal corporation has income from fees and fines.

4. *Loan*: The municipal corporation can take loan with the permission of the State Government.

19. Port Trust

In cities having port, there is port trust to look after the development of the port. The port trust is responsible for making arrangement of light, drinking water, health and sanitation of the ports and stay of the travellers. The port trust has elected members. Some members are also nominated by the Government.

20. Improvement Trust

In big cities and metropolis, there are Improvement Trust. The Improvement Trust is established by the Government of the State. It has a chairman and committee. The appointment of chairman is held by the Government. The other members are elected and nominated. The improvement trust makes plan for the expansion of roads, construction of colonies universities, hospitals, market centres, parks, bus stand, and arrange of light, latrines, drainage, drinking water, etc. It prepares greater plan for the expansion of the city in all respects. It formulates two types of plan: (i) Emergency plan and (ii) Long term plan. The emergency plan includes acquisition of land for improvement of road, street, electricity, drainage, water supply, by pass road, over bridge, railway station, bus stand, new area etc. Under Long term plan, there are plans related to discharging the waste materials and beautification of the cities. Thus, improvement

Trust plays significant role in planning the development schemes for big cities.

21. Post Office

Post office is a governmental agency which plays significant role in the development of communication. It helps us in carrying messages and goods through letters and parcels from one place to another. It also issues NSS and NSC. It carries money orders from one place to another. There is also arrangement of saving account and PPF. Thus, it plays significant role in development. In each district, there is one chief post office. Other post offices are situated up to rural areas.

22. Telephone Exchange

The Government agency BSNL has established Telephone exchanges through the length and breadth of the country to convey messages through telephones, from one place to another. It has played significant role in communication development. It has lessened the distance and time of communication not only within the country but in foreign countries, too.

23. Mobile Phones

BSNL and other private companies have provided its customers mobile phone services. Through mobile services individuals have been connected to all places of the world round the clock. This has made an individual connected to the people of entire globe.

24. Internet

BSNL and other private companies have provided its customers internet facilities. The internet functions through phones and computers. Informations from any part of the world can be sent and received with the help of internet. Internet and computer have brought information revolution in the world.

25. Co-operative Society

For the purpose of development our Government has establish cooperative societies. The cooperative societies are of various types which are as follows:

(i) Credit Co-operative Society

It is a general saying about the Indian peasants that they are borne in loan and expired in Loan. Thus, from cradle to grave they are under indebtedness. In order to make peasants free from indebtedness, our Government has established credit cooperative society. This society provides economic assistance to the peasants for agricultural works. This society provides loan to peasant at nominal rate of interest. Credit co-operative society is a village level society. It provides loan only at village level. For providing loan at district level, there is cooperative Bank.

(ii) Cooperative Agricultural Society

The main purpose of this society is to extend cooperation in agriculture and allied works. It is established for the welfare of the farmers. Co-operative agricultural society makes arrangement of loan for agriculture, irrigation, HYV seeds, fertiliser, pesticides, sale and purchase of agricultural produce and training, demonstration and exhibition for agriculture.

(iii) Industrial Co-operative Society

In order to remove poverty and provide livelihood to villagers, development of small industries, cottage industries and rural industries is a must. For this purpose industrial cooperative society is formed. This society makes arrangement of loan for establishing rice mill, chura mill, oil mill, flour mill, loaf sugar mill, weaving industry, sewing industry, leather industry, soap industry etc. It also provides loan for raw materials required for rural industries. It also organises training programme for rural industries.

(iv) Consumer Co-operative Society

This society provides loan to villagers for milch cattle, goatery, poultry, fishery, vegetable growing, cultivation of flower and fruits. It makes arrangement for the sale of those products which are used in day-to-day life in each household. It ensures the quality, weight and proper price of those commodities produced by the villagers and are consumed by the villagers.

(v) Producers Co-operative Society

This society makes arrangement to protect the interest of the producers. It lays emphasis upon community based production of goods and services.

26. Multipurpose Co-operative Society

Cooperative movements in our country began with the establishment of small societies for agriculture, rural industries, rural consumers and rural producers. These small societies had their own limitations of area, means, capital etc. These societies remained in the boundary of limited goal. To see these weaknesses, the Reserve Bank of India made advice to make credit cooperative societies multipurpose to fulfil the different needs of farmers. Multipurpose cooperative societies are meant to protect the interest of farmers, labours, exploited and deprived class and weaker sections. This society provides loan, makes arrangement of HYV seeds and fertilizer, pays attention towards health and education and encourages small industries, rural industries and cottage industries. Thus, the work of Multipurpose Co-operative Society is multidimensional. It functions in democratic manner. It has government employee, too.

27. Banks

Nationalized banks situated in rural areas as well as in urban areas are playing positive role in development process. They are providing loan and subsidy to villagers for agriculture, animal husbandry, poultry, goatery, pigery, fishery, floriculture, sericulture, beekeeping, rural industries and for small business. They are providing loan for house construction, education of children, marriage of daughter, treatment of diseases, purchasing car, truck, bus, tempoo, trekkers etc. We have now also private banks which are also extending their services in the development.

28. NABARD (National Bank of Agriculture and Rural Development)

NABARD was established in our country on 12th July of 1982. The chief office of this institution is situated at New Delhi. Its main aim is policy formulation and programme formulation for rural and agricultural development. It makes available funds for rural and agricultural development. This bank provides funds for agriculture, sericulture, horticulture, small industries, rural industries, weaving industry, sewing industry, handicraft industries and other activities related to rural development. This bank gives funds to NGOs and voluntary organisations for agricultural and rural development.

29. CAPART (Council for Advancement of People's Action and Rural Technology)

CAPART was established in the month of September 1986 in our country by the Government of India. Its chief office is situated at New Delhi. It is chief organisation for providing funds to NGOs and voluntary organisations. It provides fund for house construction, drinking water facility, health, sanitating, water harvesting, soil conservation, environmental cleanliness, etc. Since such type of developmental activities are done by NGOs and voluntary organisations, therefore, CAPART makes available funds to these organisations. CAPART makes evaluation of the developmental activities carried out by NGOs and voluntary organisations. The evaluation report of CAPART indicates that mushroom growth of NGOs and voluntary organisations has taken place. They are not social workers, rather they are professionals and money minded. They do not want to volunteer services to deprived and undeveloped people, rather they want to grasp Government funds in their own pockets in the name of NGOs and voluntary organisations. They have not will of service, rather they have will of money in their pocket. Therefore, negative trend of making money and professionalism behind opening NGOs be discouraged. It also reveals corruption prevalent in granting funds to NGOs. Top politicians and bureaucrats are involved in taking bribes and commission from NGOs for providing funds for development activities. They are also involved in presenting NGOs to their daughters, daughter-in-laws; sons, son-in-laws and other nearest relatives.

30. LAMPS (Large Area Multipurpose Co-operative Societies)

Lamps were organised in tribal areas under tribal sub-plan approach during Fifth Plan period. Lamps were organised to fulfil the following objectives:

(i) To provide consumption and production credit to tribals
(ii) To purchase surplus farm and forest produce
(iii) To meet consumers need and farm inputs.

LAMPS were set up either by converting the existing and working primary agricultural cooperative societies (PACS) or by organising new ones at panchayat level or Block level. LAMPS were set up as an alternative to traders so far as procurement and marketing of MFP (Minor Forest Produce) was concerned. It is ground level agency for collection, procurement and marketing of MFP.

31. State Forest Development Corporation/Tribal Development Co-operative Corporations/Trade and Development Federation

These are state level agencies known by different names in different states for the collection, procurement and marketing of MFP. They have their own collection centres LAMPS employ private contractors for collection of MFP from primary collectors, i.e. tribals. These state agencies organise only nationalised items of MFP. They are also responsible for storing, processing and marketing of MFP. They fix collection price and make payment to primary collectors. The main purpose behind establishing these state level Governmental agency was to protect the tribals from the exploitation of *mahajan*, traders or local contractors.

32. TRIFED (Tribal Co-operative Marketing Development Federation of India Limited)

This is national level Government agencies involved in collection, storage, processing and marketing of MFP. It was established on 6th August, 1987 under Multistate Co-operative Society Act, 1984 by the Ministry of Welfare, Government of India. The prime objectives of the TRIFED are as follows:

(i) To ensure remunatrative prices for tribal produce—forest and agricultural
(ii) To provide marketing infrastructural facilities for both national and international sales
(iii) Value addition of tribal produce in tribal area as far as possible.

To achieve its objective TRIFED has established a wide network by setting up four zonal offices, nine regional offices and four field offices.

33. National Scheduled Castes and Scheduled Tribes Finance and Development Corporation

Through this institution a group of SCs and STs can take loan ranging from 35,000 to 10 lakh. 25 per cent of the amount is subsidy. On remaining amount interest is to be paid as decided by the institution from time to time. Auto rickshaws, buses, trucks, jeeps, Trekkers, trucks can be obtained by a group of SCs or STs through this institution

34. Ministry of Home Affairs

For development of a nation and its people, situation of peace and harmony is a must. Maintaining peace and harmony in country is responsibility of Home Ministry of the Government of India as well as the Government of the States. Since last three and half decades our country is facing the problem of terrorism and extremism. But the efforts made by Ministry of Home Affairs of our country and states, the pace of development has been maintained in our country. We are marching ahead on different parameters of development.

The matters concerning tribal welfare and development were also dealt by a DIVISION of Home Affairs known as Tribal DIVISION after independence upto September 1985.

35. Ministry of Human Resource Development

The Ministry of Human Resource Development is responsible for educational development in the country. Previously this ministry was known as Ministry of Education, but now its name has been changed as Human Resource Development Ministry. Our Central as well as State Government have this ministry for educational development in the country and states. This ministry lays emphasis upon the establishment of schools, colleges, universities, technical institutes, management institutes etc. Educational policies, plans and programmes are formulated and implemented by this ministry.

36. Ministry of Science and Technology

This ministry of the Government of our country is responsible for the development of science and technology in our country. Today we are living in era of science and technology. This era is also known as era of knowledge revolution or information revolution. Knowledge revolution has been brought by the advancement of science and technology. Today the powerful is one who possesses knowledge. Those who lack knowledge are powerless. In this race of knowledge based power, it is the responsibility of the Ministry of Science and Technology to enhance the knowledge of our people in different areas through latest technological advancement and imparting knowledge in that direction.

37. Ministry of Welfare

The Ministry of Welfare was created in September 1985 from the Ministry of Home Affairs. This ministry formulates policies regarding the welfare programmes. It also makes available funds to different organisations for welfare activities. At state level, there is welfare ministry, too. In tribal concentrated states, the state has independent Tribal Welfare Ministry. The Welfare Ministry is advised by two bodies—(i) Tribe Advisory Council and (ii) Tribal Research Institute. These two bodies extend advice to ministry to welfare in framing policies and programmes of tribal welfare. A Deputy Director for Backward Classes is posted by the centre who acts as liaison between the national and state level welfare activities and development of tribals.

38. Ministry of Social Justice and Empowerment

Ministry of Social Justice and Empowerment was created in May 1998. This Ministry looks after social justice and empowerment of Scheduled Tribes, Scheduled Castes, Backward Classes, Women, Children, Aged etc. This Ministry protects the interest of the weaker sections of the society and makes arrangement of their empowerment by organising them through Non-Governmental Organisations. This Ministry has appointed several commissions for justice and empowerment of Weaker Section and deprived communities of our country.

39. Ministry of Tribal Affairs

The Ministry of Tribal Affairs was constituted in October 1999 by bifurcation of Ministry of Social Justice and Empowerment. The main objective behind establishing this ministry was to focus attention on integrated socio-economic development of Scheduled Tribes in a coordinated and planned manner. Before the creation of Ministry of Tribal Affairs, matters concerning Tribal Affairs and development were dealt at Government of India level by the ministries as named below at different points of time:

(i) As division of Ministry of Home Affairs known as Tribal Affairs after Independence to September 1985.
(ii) Ministry of Welfare from September 1985 to May 1998.
(iii) Ministry of Social Justice and Empowerment from May 1998 to September 1999.

The Ministry of Tribal Affairs started functioning in October 1999 with the schemes/programmes dealt erstwhile by Tribal Division of Ministry of Social Justice and Empowerment and continued with those schemes.

The process of the constitution of National level Tribal Development and Finance Corporation by bifurcation of National Scheduled Castes and Scheduled Tribes Finance and Development Corporation (NSFDC) of the Ministry of the Social Justice and Empowerment has been completed. It made possible for the new corporation to commence its concentrated operation on economic development of Scheduled Tribes from April 2001.

Mandate of the Ministry: The Ministry of Tribal Affairs has the following mandate:

(i) Tribal welfare planning, policy formulation, research and training.
(ii) Tribal Development including scholarships to STs.
(iii) Promotion of voluntary efforts in development of STs.
(iv) Administrative ministry with respect to matters concerning:

- Scheduled Area
- Autonomous districts of Assam excluding roads, bridge works and ferries thereon.
- Regulations framed by Governors of state for Scheduled Areas and for Tribal Areas specified in the sixth schedule to the Constitution

- National Commission for Scheduled Castes and Scheduled Tribes in so far as they relate to STs, and
- Issue of directions regarding the drawing up and execution of schemes essential for the welfare of the Scheduled Tribes.

The Ministry of Tribal Affairs is the nodal ministry for overall policy, planning and coordination of programmes of development for Scheduled Tribes.

Recent Initiatives of the Ministry: For more focus on Tribal Welfare and Development, the Ministry of Tribal Affairs have taken up following initiatives:

(i) Direction of states for the adoption of Maharashtra pattern of planning and devolution of funds earmarked for tribal sub plan (TSP) areas. In Maharastra, the funds earmarked for TSP are Tribal Development Department of State Government which releases funds to different department for execution of development works under TSP. This process ensures better utilisation of TSP funds for tribal areas.

(ii) Introduction of a new central sector scheme exclusively for development of primitive tribal groups, the most underdeveloped group among the tribals.

(iii) Promotion of more Non-Governmental Organisations (N.G.Os) in the field of tribal development.

(iv) Enforcement of strict monitoring mechanism on performance of NGOs through officers of ministry and also by involvement of State Government mechanism machinery to identify the non-performing NGOs and weeding them out.

(v) Indentifying non-Governmental organisations which have an all India character and known for their selfless character and service. Such identified NGOs be known as established voluntary agencies.

(vi) Setting up of a National Institute for Research and Training in Tribal Affairs (NITA) as apex body of State level Tribal Research Institute.

(vii) Revision of schemes raising the cost heads to realistic and workable limits.

(viii) Enlistment of more work programmes/activities mostly on income generation and skill development to the list of projects for Grant-in-aid to NGOs and also making provisions for cent percent project cost as grants in most deserving cases.

(ix) Enhancement of rates of fellowships and grants for supporting projects, seminars and publications.

***Plans and Programmes of the Ministry*:** The Ministry of Tribal Affairs continued to implement various schemes/ programmes aimed at welfare and development of Scheduled Tribes. The plans and programmes of the Ministry are as follows:

(1) Central sector schemes under which 100 per cent assistance is given to States/UTs and Government of India undertakings.

These Central sector schemes are as follows:

- Special Central Assistance (SCA) to Tribal Sub-Plan (TSP).
- Grants under First provision to Article 275U of the Constitution.
- Grant-in-Aid to State Tribal Development Corporations (STDCs).
- Price support operations to TRIFED
- Investment in share capital of Tribal Cooperative Marketing Development Federation of India Limited (TRIFED).
- Establishment of Grain Banks in Tribal Villages.

(2) Centrally sponsored schemes matching Assistance (50:50) is given to States/UTS).

These centrally sponsored schemes are as follows:

- Construction of Girls Hostels for Scheduled Tribes
- Construction of Boys Hostels for Scheduled Tribes

- Establishment of Ashram Schools in Tribal Sub Plan Area
- Research and Training

(3) Central sector schemes under which 100 per cent assistance is given to States/UTS and NGOs
These schemes are as follows:

- Vocational Training in Tribal Area
- Educational complex in low literacy pockets for development of women literacy in Tribal Area
- Development of primitive Tribal Groups (PTGs)

(4) Central schemes under which Ninety to cent-per-cent assistance is given to voluntary organisations.
These schemes include Grant-in-aid to voluntary organisations working for the welfare of the Scheduled Tribes

(5) Central sector/Centrally sponsored schemes bifurcated from the Ministry of Social Justice and Empowerment.
These schemes are as follows:

- Post matric scholarships to students belonging to Scheduled Tribes for study in India.
- National overseas scholarship for STs candidates
- Book Banks for Scheduled Tribes students
- Coaching and allied scheme for scheduled Tribes students
- Assistance to State Scheduled Tribe Development Corporation (STDCS)
- Central Sector Scheme of upgradation of merits of Scheduled Tribes students.

40. Tribal Sub-plan

The tribal sub-plan strategy was evolved by an expert committee setup by the ministry of Education and social welfare in 1972 under the chairmanship of professor S.C. Dube for rapid socio-economic development of tribals. The TSP is a plan within the ambit of a state or union Territory. Plan meant for welfare and development of tribals. This plan is a part of overall plan of a State. Therefore, it is termed as sub-plan. The benefits percolated to the tribals and tribal areas of a State/UT from TSP are in addition to what percolates from the overall plan of a State/UT

The TSP is in operation in 21 States and 2 UTs. In the states like Arunachal Pradesh, Meghalaya, Mizoram and Nagaland, the TSP concept is non-applicable. Since in these States tribals represent more than 80 per cent of the state population.

The tribal sub-plan identifies the resources for TSP areas. It prepares a broad policy framework for the development and it defines a suitable administrative strategy for its implementation.

The most significant aspect of this strategy is to ensure flow of funds for TSP area at least in equal proportion of Scheduled Tribes population of each State and UT.

The twin objectives of TSP approach are:

(i) Over all socio-economic development of tribals and to raise them also above poverty level.
(ii) Protection of tribals from various forms of exploitation.

The coverage of tribal population under TSP of 23 States and UTs is almost total at present under one of the following depending upon their population percentile:

(i) Integrated Area Development Projects (IADPs)
(ii) Modified Area Development Approach (MADA)
(iii) Clusters
(iv) Primetive Tribal Groups (PTGs)

TSP ensures allocation of fund for tribal areas from state plan as well central ministries at least proportionate to population of tribals in State (state plan) and to the overall proportionate tribal population for the country from the budget of Central ministries and financial institutions.

41. Backward Class Commission

The Government of India constituted a commission to identify backward classes and to protect the interests of Scheduled Castes and Scheduled Tribes. Thus, in the beginning Scheduled Tribes and Scheduled Castes were brought under the Backward Classes Commission along with backward castes. This Commission was constituted in 1953 under the chairmanship of Kaka Kalelkar. This Commission functioned till 1955. The Commission in its report submitted to the Government in 1955 recommended for the welfare and development of backward classes, i.e., the weaker sections of the society.

42. Commissioner for Scheduled Castes and Scheduled Tribes

As per Article 339 of Constitution of India, capital there is provision to constitute a commission for Scheduled Castes and Scheduled Tribes by Hon'ble President, of country. The special officer of this commission is known as Commissioner for Scheduled Tribes and Scheduled Castes Commission. This commission was constituted in 1958-59 and continued till 1986-87. The Commissioner for Scheduled Tribes and Scheduled Castes had to submit an annual report regarding administration and welfare of Scheduled Tribes and Scheduled Castes to be tabled in the parliament for detail discussion B.D. Sharma was last Commissioner for SCs and STs.

43. National Commission for Scheduled Castes and Scheduled Tribes

As per 65th amendment to Article 338 of Constitution of India, there is provision to constitute a national commission for Scheduled Castes and scheduled Tribes by Hon'ble President of our country. This national level commission for SCs and S.Ts. can investigate, monitor and evaluate all matters related to the constitutional safeguards of SCs and STs. This commission was constituted in September 1987. The commission has sixteen state offices at different state head-quarters. The commission has a chairman, a deputy chairman and five members. The commission has to submit an annual report to the President of India on the safeguards of constitutional provisions, development and welfare of SCs and STs

The commission has been reviewing the working of the constitutional safeguards and implementation of various development programmes for SCs and STs in state level review meeting and through the visit of the commission and its members. State offices are eyes and ears of the commission. The commission ensures the utilisation of funds in TSP area as per the guidelines of the Planning Commission. It also ensures the utilisation of special Central assistance received from the Central Government on basic infrastructure and minimum requirements such as roads, electricity, house, drinking water, health, nutrition hygiene and sanitation. It also examines whether the exploitation and oppression of tribals have been ended or not. It also examines the cases related to atrocities on SCs and STs. It is the bounden duty of the commission to prevent atrocities on SCs and STs and to ensure relief and rehabilitation of victims of such offences.

44. National Scheduled Tribes Commission

As per Article 339 of Constitution of India, there is provison to constitute a commission by Hon'ble President of our country to report on the administration of the scheduled areas and the welfare of Scheduled Tribes in the states. According to this, Scheduled Castes Commission and Scheduled Tribes Commission has been bifurcated.

45. Other Backward Classes (OBC) Commission

As per Article 340 of Constitution of India, there is provision to appoint a commission to investigate the conditions of socially and educationally backward classes and the difficulties under which they labour and to make recommendations to remove such difficulty and to improve their conditions.

Previously only SCs and STs were given the benefit of 22.5 per cent reservation in Central Government jobs. A 27 per cent reservation for OBCs in Central Government jobs was introduced on the recommendation of the report of Mandal Commission. The demand for the reservation of seats in Government jobs by OBCs got momentum in early seventies. When Janata Government came to power in 1977, B.P. Mandal Commission was constituted in 1979. The task of this Commission was to identify OBCs and to suggest percentage of reservation in the jobs for OBCs. The Commission submitted its reports in 1980. The Commission in its report submitted to the Government in 1980 suggested for 27 per cent reservation to OBCs in Government jobs. When Janata Dal Government came to power in 1989, declaration for the implementation of Mandal commission's report was made. This led to all over India agitation by the students belonging to FCs (Forward Classes). Several students made suicide by burning in protest against the decision to implement Mandal Commission's report. This decision could not be implemented due to fall of V.P. Singh Government at Centre. Ultimately Mandal Commission report was implemented in 1993 during Narasimha Rao Government with a concept of creamy layer not to get this benefit.

OBCs have been divided politically into two categories: (i) Annexture-I and (ii) Annexture-II. In Annexture-I those OBCs have been included who are socially and educationally less developed. Such as Malah, Nonia, Barhi, Kumhar, Kahar etc. But in Annexture-II those OBCs have been placed who are socially and educationally more well to do such as Yadav, Koeri, Kurmi, Bania etc. There are 3920 OBC communities in our country. OBCs commission exists at central as well as state level. Recently Manmohan Singh Government has introduced 27 per cent reservations to OBCs students in IITs and All India Management Institutes.

46. Planning Commission

Planning Commission in our country was constituted in 1950. Since then it is functioning. The prime minister is chairman of this commission. Basides prime minister as chairman, this commission has a depty chairman and mempers. The Planning Commission transacts the following business:

(i) To evaluate physical capital and human resources.
(ii) To formulate policy and plan for balanced utility of physical capital and human resources.
(iii) To determine the stages of plan and to arrange resources for each state of plan.
(vi) To evaluate the factors responsible for creating hindrance in growth rate.
(v) To formulate administrative design for implementation of plans.
(vi) To evaluate the implementation and monitoring of the plans.

Now we are under 11th Plan. 10th Plan has end in March 2007. So, attempts have been made to formulate the strategy of 11th Plan period.

47. Women Commission

National Women Commission was constituted during 8th Five Years Plan period under the National Commission for Women Act, 1990. It is a statutory body. It is functioning since 1992. Its main duties are the protection of interests of women by safeguarding their rights.

48. Rashtriya Mahila Kosh

Rastriya Mahila Kosh was set up in 1993 mainly to facilitate the credit support and microfinance to poor women.

49. Human Rights Commission

Now-a-days, Human Rights Commission is an international truth. As per declaration of Human Rights (1948), there is provision for constitution of Human Rights Commission to safeguard the Human Rights of children, women, aged, labourers and indigenous people. We have International

Human Rights Commission. But in our country the office of Human Rights Commission is at New Delhi and in each states.

50. Parliamentary Committee on Welfare Activities of SCs and STs

Our parliament also has indirect control over the welfare activities of SCs and STs. Its functions are as follows:

(i) To consider the reports submitted by Commissioner for SCs and STs, under Article 338(2) of Constitution and to report the both houses the measure that should be taken by the Union Government including the administration of UTS.
(ii) To report both houses on action taken by the Union Government and the administration of UTs, on the measures already proposed by the committee.
(iii) To examine the measures taken by Union Government to secure due representation of SCs and STs in service and posts under its control.
(iv) To report both houses on working of welfare programmes on SCs and STs of Union Territories.
(v) To report both houses on all matters concerning the welfare of SCs and STs.
(vi) To examine such matters as may seen fit to committee or are referred to it by the House or speaker.

From time to time our Government has constituted different commissions to assess and analyse the welfare works. Some of these committees are as follows:

(1) The criminal Tribes Act Enquiry Committee (1949-50). The chairman of this committee was Anantha Sayanam Ayyangar.
(2) Backward Classes Commission (1953-55). The chairman of this committee was Kaka Kalelkar.
(3) Study Team on social welfare and welfare of Backward classes (1958-59). The chairperson of this study Team was Renuka Ray.

(4) Committee on special multipurpose Tribal Blocks (1959-60). The chairman was Verrier Elwin.

(5) Scavenging Conditions Enquiry Committee (1957-60). The chairman was N.R. Malkani.

(6) Scheduled Areas and Scheduled Tribes Commission (1960-61). U.N. Dhebar was chairman of this committee.

(7) Study Group on welfare of weaker sections of the village community (1960-61). Jaya Prakash Narayan was chairman.

(8) Special Working Groups on Cooperation of Backward Classes (1961-62). The chairman was M.P. Bhargava.

(9) Advisory committee on revision of list of STs and SCs (1965). Chairman was B.N. Lokur.

(10) Committee on Customary Rights of Scavengers (1965-66). Chairman was N.R. Malkani.

(11.) Committee on Tribal Economy of Forest Areas (1965-67). The chairman was Hari Singh.

(12) Working Group to study the progress of measures for land allotment to Scheduled Castes and their representation in services (1967). The chairman was M.R. Yardi.

(13) Committee on Untouchability, Economic and Educational Development of Scheduled Castes (1965-69). The chairman was Elayaperumal.

(14) Tribal Sub-plan Strategy Expert Committee (1972). The chairman of this expert committee was S.C. Dube.

(15) Study Group on Tribal Labour (1966-69). The chairman was L.P. Vidyarthi.

(16) Task Force on Development of Tribal Areas by Planning Commission (1972-73). Chairman was L.P. Vidyarthi.

(17) Study Team of Tribal Development Programmes (1966-69). Chairman was P. Shilu Ao.

51. Tribal Research Institutes

In order to facilitate the development and welfare works, the welfare department of State Government has two bodies—(i) Tribal Research Institutes and (ii) Tribal Advisory Council.

In ,major tribal concentrated areas. The State has an independent Tribal welfare ministry. The state tribal welfare ministry is advised by these two bodies in formulating policies and plans of Tribal welfare and development. Thus Tribal Research Institute is an organ of state Ministry of Tribal Welfare in tribal concentrated states. In course of time the following Tribal Research Institutes have been established in our country:

1. Tribal Welfare Research Institute, Chhindwara Chhattisgarh
2. Tribal Welfare Research Institute, Ranchi, Jharkhand
3. Tribal Welfare Research Bureau, Bhubaneshwar, Orissa
4. Tribal Welfare Research Institute, Kolkata, W.B.
5. Tribal Welfare Research Institute, Shillong, Meghalaya
6. Tribal Welfare Research Institute, Udaypur, Rajasthan
7. Tribal Welfare Research Institute, Hyderabad, Andhra Pradesh
8. Tribal Research Institute, Lucknow, U.P.
9. Tribal Research Institute, Poona (Maharashtra)
10. Tribal Research Institute, Ahmedabad (Gujarat)
11. Tribal Research Institute, Chennai (Tamil Nadu)
12. Tribal Research Institute, Imphal, Manipur
13. Tribal Research Institute, Aggrtala (Tripura)
14. Tribal Research Institute, Trivendrum (Kerala)
15. Tribal Welfare Research Institute, Poona (Maharashtra)
16. Tribal Welfare Research Institute, Gujarat Vidyapeeth, Gujarat.

These Tribal Research Institutes make ethnographic and problem oriented study of tribal communities located in their jurisdiction. They suggest the ministry of welfare, Government of State to formulate policies and programmes of tribal welfare and development.

52. Tribal Advisory Council (TAC)

As referred to above, the State Ministry of Tribal Welfare has two bodies to provide suggestions regarding the welfare and development of the tribals living in the state. One body is Tribal Research Institute and the other is Tribal Advisory Council. Thus, Tribal Advisory Council is a body to provide suggestions to State Tribal Welfare Ministry for formulating and implementing policies and programmes of tribal welfare and development works.

According to section 4 of the 5th schedule to Constitution of India, there is provision for constitution of Tribal Advisory Council in all states having Scheduled areas. TAC can be constituted in those states, too. Which do not have Scheduled Areas, but have tribal concentration. In the States having Scheduled areas like Rajasthan, Jharkhand along with Bihar, Madhya Pradesh along with Chhattisgarh, Rajasthan, Orissa, Maharashtra and Himachal Pradesh have Tribal Advisory Council. TAC has also been constituted in states like West Bengal and Tamil Nadu, which have tribal population but not Scheduled areas. TAC must have 25 members. Three-forth members must be represented by MLAs belonging to STs. The rest seats are filled up by other kind of members belonging to STs. In TAC some persons having extra ordinary knowledge of tribal culture and problems can also be co-opted as its member. Governor can formulate the rule governing the TAC. The Governor may suggest TAC to take steps for the welfare of the tribals. Our Constitution has provided enormous power to TAC for the development and welfare of the tribals.

From the above account it is clear that TAC is a body appointed by the State Government as per rules framed by the Governor and as per the provision of the 5th schedule of Constitution of India. It monitors and reviews the measures to be taken for the welfare and advancement of tribals in the state. It makes advice to the Governor on that basis. It has MLAs, secretary, and Tribal Development Commissioner. Tribal Welfare Minister is chairman of this council. Council meeting is held once in every three months. The main objective behind the constitution of TAC is to prevention of tribals from exploitation and proper utilisation of special central assistance and TSP funds in the interest of scheduled areas and tribal people.

53. Autonomous District Council and Regional Council

As per the sub-para 6 of para 2 to 6th Schedule to Constitution of India, the Governor of the states having tribal areas can formulate autonomous district council and antonomous regional council for administration of tribal area. In four states of India, there are tribal areas. These states are Assam, Meghalaya, Mizoram and Tripura. In all these four states, there are nine autonomous districts. The states of Meghalaya, Mizoram, Assam and Tripura have three, two and one autonomous districts respectively. Generally an autonomous district is represented by a particular tribe. But if there are more than one tribe in an autonomous district, then the Governor can notify the area as Autonomous Region even within the autonomous district. But the situation in Tripura is different. In Tripura, an autonomous district has three revenue districts having different tribes. The Governor can increase or decrease the boundary of an autonomous district by issuing notification.

Generally, in each autonomous district there is an autonomous district council for the purpose of administration. The total members of an autonomous district council is not more than 30. The Governor nominates four members out of total 30 members. 26 members are elected through adult franchise. An autonomous district council may have several regional constituencies. The Governor determines the number of district or regional council. The Governor may reserve all constituencies for tribal candidates only. The Governor may prohibit a non-tribal candidate from seeking election. District council is responsible for formulation of policies for the development of tribal people living within the boundary of an autonomous council. It also transacts various other kinds of business and activities.

District council/Regional council has been given power to formulate the following kind of rules:

1. Rule related to distribution, acquisition and use of land other than reserved forests.
2. Rule regarding the management of all types of forests.

3. Rule regarding the development of canal or other water sources for the purpose of irrigation.
4. Rules regarding ban on Jhum or shifting cultivation.
5. Rules related to constitution of Gram samiti, Nagar samiti, boards and councils.
6. Rules related to the development of village police, town police, education, health, sanitation etc.
7. Rules regarding the appointment of Mukhia or Pramukh.
8. Rules regarding the inheritance.
9. Rules regarding marriage and divorce.
10. Rules related to social customs.
11. Rules to constitute village court.

The council also performs functions related to schools, hospitals, road, fishery, road transport, water transport etc.

Each autonomous District council/Regional council has its own fund. The Governor may formulate rule for getting fund and making expenditure. The council may collect the following kind of taxes:

1. Taxes from traders and businessmen
2. Taxs on animal transport, road transport and water transport.
3. Sale and purchase of commodities in the market.
4. School tax, road tax, hospital tax, water tax, latrine tax, etc.

The council may formulate the following laws to check the exploitation of tribals

(i) To issue licence to selected sahukars for doing money lending and a complete ban on unlicensed money lenders.
(ii) To determine the rate of interest.
(iii) To review the account register of licensed money lenders.
(iv) Time to time renewal of the license of money lenders.

The Governor may appoint a commission to make inquiry into administrative functioning of the ADC. The report of the commission is tabled in the Assembly. The Governor can dissolve the ADC/ARC on the recommendation of the commission.

54. Anthropological Survey of India

Anthropological Survey of India with its headquarters at Kolkata started functioning since 1946. It has its sub stations at Shillong, Dehradun, Bhopal, Nagpur, Mysore, Port Blair etc. Although, it is not directly related to the development, but by conducting social, cultural, psychological and physical studies of the people, it has contributed much towards the understanding of different people and their problems. Thus, anthropological survey of India has been helpful to welfare and tribal development.

52. Social-Cultural Research Institutes

In our country state governments have set up Social Cultural Research Institutes. The main purpose of these research institutes are to record the ethnography of the community and to highlight their problems so that government could take preparation to eradicate them.

53. Harijan Research Institutes

In Scheduled Castes dominated states Harijan Cells and Harijan Research Institutes have been set up by the Government to study the problem of Harijan exploitation, oppression, atrocities, compensation and rehabilitation. These institutes makes suggestions to the Government regarding their welfare and development.

8

Non-Governmental Agencies of Development

Development is not a simple phenomenon. Rather, it is a complex phenomenon. It is true that to bring all people in the main stream of development is the prime duty of the Government. But it is also true that Government alone cannot do this complex job. The development of weaker sections cannot be achieved through the Governmental Agencies alone. It is because Governmental Agencies have their own limitations. Several evaluation studies of on going development programmes revealed that only partial advantages researched to the real beneficiaries, while major portion of benefit went in the pockets of the concerned Governmental Agencies. The reports also revealed that the Governmental Agencies could not understand the importance of cultural values and traditions of the local people. They imposed a number of developmental programmes on the heads of local people without knowing their need and value. Therefore, those programmes were not accepted by the people. The Governmental agencies were not able to establish good rapport with the local people, therefore, they were treated as outsiders. They were not accepted as development agent by local people. In fact Governmental agencies could not win the confidence of the people in implementing the programme of development. Large scale prevalence of corruption in Governmental Agencies also came to knowledge through evaluation reports.

Considering the recommendations of the evaluation study reports, our Government also understood that the strategy for rapid development of Scheduled Castes, Scheduled Tribes and other weaker sections of the society requires through understanding of their socio-economic values, tradition and culture. There is no doubt that sometimes the development programmes do suffer from lack of knowledge of Government officials of tribal values, customs, traditions and cultures.

Keeping these facts in mind, our Government decided to promote voluntary efforts in development of SCs and STs. Our Government has promoted voluntary efforts in different areas of development such as literacy, education, drinking water, irrigation, water harvesting soil conservation, forest conservation, house construction, vocational training, preservation of human rights, preservation of cultural and natural heritage, health nutrition, sanitation, legal redressal services, need based socio-economic upliftment, case of orphans, handicapped and mentally retarded, social justice and empowerment of children, women, aged, SCs, STs and indigenous people.

A number of voluntary organisations have been rendering valuable services to the people by living amidst them. The people are also found more open with them. Voluntary organisations are valuable non-Governmental Agencies of development. Therefore, they are known as Non-voluntary organisations (NGOs). They have attracted the attention of our Government in the field of development due to the following reasons:

1. Failure of development programmes run by the Governmental Agencies
2. Financial irregularities done by the Governmental Agencies
3. Not completion of development schemes in time by the Governmental Agencies
4. Oppression of beneficiaries by Governmental Agencies
5. Not proper selection of beneficiaries by the Governmental Agencies
6. Lack of coordination between the local people and the Governmental Agencies
7. Lack of cultural knowledge to Governmental Agencies

8. To make the implementation of development programme a complex phenomenon by the Governmental Agencies.
9. Large scale corruption prevalent in Governmental Agencies
10. Lack of commitment, honesty and sincerity in Governmental Agencies
11. Presence of sentiment of service, honesty, sincerity and commitment in voluntary organisations
12. In order to provide employment to unemployed youth through voluntary organisations.

Voluntary Organisations and Non-Governmental Organisation

It will be the fitness of the thing to make a conceptual difference between Voluntary Organisation and Non-Governmental Organisation. There is no doubt that Non-Governmental Organisation is also a voluntary organisation. But not all voluntary organisations are Non-Governmental Organisation. Only those voluntary organisations are non-Governmental Organisations who have registered themselves under the Societies Registration Act, 1960 or any other statute of centre or the state including public trust, company, educational and other institutions, local bodies or cooperative societies having sound financial position and aptitude and experience in welfare activities. There are also a number of voluntary organisations which are formed for specific purposes without any registration to Government. For example, voluntary organisation formed on the occasion of worship, fair, festivals, pilgrimage etc. Such voluntary organisation come into existence for specific period. They are dissolved after the purpose is over. They do not have permanent registration number. Therefore, they do not come under the category of NGOs.

Chief Characteristics of NGOs

The chief characteristics of an NGOs are as follows:

1. It considers the service of man as its great religion

2. Its members have disciplined life
3. No feeling of high-low and rich-poor among members
4. Greater sense of cooperation among members
5. Simple life of members
6. Prevalence of sense of service and humanity among members
7. Rendering services by members on honorarium or nominal remuneration
8. Use of kinship terminologies for members and beneficiaries
9. Dedication and commitment of members towards work
10. Emphasis on timely completion of work
11. Strict action against indiscipline and financial irregularity
12. Proper coordination between people and the government
13. Knowledge of cultural values of the people for whom they work or want to work
14. Time to time discussion with members to remove errors and faults
15. Financially sound
16. Must possess permanent Registration number of the government.

Structure and Administration of NGOs

For any NGOs, structure and administration are essential and integrated parts. The structure and administration of NGOs , reveal the following facts:

1. Each NGO has a president. The selection of the president is done on consensus basis among members of NGO.
2. There is a secretary to provide suggestion to the president of the NGO. He/She is elected through secret balot papers.
3. There is treasurer to keep account of funds received by NGO from the government for development work. He/she is also elected.

4. There are executive members who are elected by the people. The number of executive members depends upon the size and the area of NGO.
5. The meeting of officials of NGO takes place at regular intervals to discuss the problems and to review the progress of the work going on.
6. In order to get funds from the government, registration number of the Government is a must. Without registration of the Government, NGO cannot receive funds from the Government.
7. NGO has to fulfil the conditions laid down by the Government, otherwise its registration number can be cancelled.
8. Each NGO has some workers who receive some nominal remuneration to make a living.
9. The post of president, secretary, treasurer and executive members are honorary.
10. Each registered NGO has an annual budget. The statement of expenditure of funds is to be sent to the Government.
11. Governmental Agencies provide monitoring to NGOs. They have control over NGO and make evaluation of work performed by them.
12. The NGO has to seek support of the local Government administration.
13. It is the duty of president to establish contact with government officials. He/She makes the government officials aware of work done by the NGO.
14. The president also works as a connecting link between the government official and the local people or beneficiaries.
15. The president acts democratically. He is all in all of the NGO. He can oust any member any time. He does not take any decision individually. He takes decision with the help of his associates.
16. Some NGOs have local status, some NGOs have regional status, some NGOs have state status and some have national statues.
17. There are some International level NGO, too.

18. National and International level NGOs have bigger organisation. They can do national and international level work.

NGOs and Government Polices

Our Government formulates policies regarding the allotment and grant of funds to NGOs. The Government formulates budget of work to be done by NGOs. The fund is given to the extent of 100 per cent directly to NGOs working in Scheduled areas on the recommendations of the concerned State Government/UT Administration. The NGOs are supported to the extent of 90 per cent or the project cost for the projects in areas other than Scheduled Areas. However, the concerned ministry has discretion to assist even up to 100 per cent in case of Established NGOs. The ministry may grant assistance even without recommendation of State Government/UT Administration.

Development Work Assigned to NGO

The Grants are provided to the NGOs on applications in prescribed formats supporting the projects and meeting the objectives of the scheme. Project types for receiving grant-in-aid by NGOs form the ministry of tribal Affairs are as follows:

Sl.No.	*Type of Project*
1.	Residential school for 100 students
2.	Non-residential school for 100 students
3.	Construction of hostels for 100 students
4.	Banasthali Vidyapith, Banasthali, Rajasthan for 15 Tribal Girls students each from Assam, Meghalaya, Mizoram, Manipur, Nagaland, Tripura, Arunachal Pradesh, Sikkim, Andaman and Nicobar Islands and Lakshdweep
5.	Rural Night School for tribals. 50 students per school for 50 schools.
6.	Library
7.	Mobile Library-cum-Audio visual unit
8.	Ten bedded hospital/dispensory

(contd.)

(contd.)

Sl.No.	Type of Project
9.	Mobile Dispensary.
10.	Centre for Training on computer (30 trainees)
11.	Typing and short hand training on computer (30 trainees)
12.	Divyayan: Training in Agriculture and allied subjects for 100 tribals residential and follow up programmes R.K. Mission Merabadi Ranchi
13.	Training in carpentry (40 trainees)
14.	Workers Training centres, (Residential, 25 Trainees)
15.	Training centre for Dari making (50 Trainees)
16.	TV, VCR, Tape Recorder, Radio/Transistor Training centre (40 Trainees)
17.	Cane and Bamboo training centre (25 trainees)
18.	Coconut coir making and Agarbati making centre (60 Trainees)
19.	Training centre for composing, printing and book binding (60 trainees)
20.	Training to welding and fitter trades (40 trainees)
21.	Knitting, weaving and handloom training centre (75 trainees)
22.	Scooter/Motor cycle repair training centre (50 trainees)
23.	Balwadi or creche centres
24.	Drinking water programmes for Tribals
25.	Environment protection programme
26.	Legal redressal services
27.	Health and sanitation programme

Allocation of Funds to NGOs

There is continuous increase in the allocation of funds to the NGOs by the Government. In the financial year 1993-94, a sum of Rs 7.50 crores was given to 208 NGOs as Grants-in-aid by the Government. In the financial year 1994-95, a sum of Rs. 11.60 crores was given to 313 NGOs as Grant-in-aid. In the financial year 1995-96, a sum of Rs. 20.33 crores was given to NGOs as Grants-in-aid.

The allocation of funds for NGOs on account of Tribal welfare and development by the Ministry of Tribal Affairs was also increased over the successive plan periods. In 9th Plan period, allocation to NGOs was 92.09 crores.

Origin and Development of NGOs in India

When our country achieved independence, social wokers working for the weaker sections got political and administrative environment very congenial. Since a number of their colleagues has entered in politics and Government, the Government sought their cooperation in the welfare and upliftment of weaker sections. Thus, social workers began to act as semi-officials and social scientists in the matter of welfare and development of STs and SCs. They began to make recommendations to the Government on the development issues. They also started getting substantial grant from the Government for the welfare and upliftment of STs and SCs. They began to work on small remuneration or honorario. They began to organise conferences at all India level. These activities of social workers gradually gave rise to voluntary organisations or NGOs in our country. As a result, a number of NGOs began to work for the betterment and upliftment of Tribals, Scheduled castes and other Deprived communities. A brief description about some NGOs is being presented below.

1. Bhartiya Adimjati Sewak Sangh

This national level NGO was started in 1948 under the presidentship of Late Dr. Rajendra Prasad, the then President of India. In the beginning, only 18 organisations began to work under this national level NGO. In the year 1950, the number of organisations working under this federation rose to 50. At present 100 organisations are affiliated to it. This national level NGOs cover all developmental issues related to tribal life. This NGO publishes news letters, magazines, pamphlets, journals; articles, books etc.

Dealing with informations on welfare and development activities of tribals, besides doing welfare and development activities for them, its members also organise and participate in national and international level seminars.

2. *Bhartiya Depressed Classes League*

This is a Delhi based NGO involoved in welfare and development activities of SCs and STs. This NGO works to eradicate the problem of untouchability, exploitation and oppression of SCs and STs. It also extends its services in literacy, education, health and sanitation development of STs, SCs and other deprived communities.

3. *Indian Red-Cross Society*

This NGO is an International NGO working in different countries of the world. The headquarters of Indian Red Cross Society is located at Delhi. Its organisations are spread throughout the length and breadth of our country. This NGO runs development and welfare programmes related to women and children such as nutrition, health, treatment of disease and sanitation. It organises camps on diseases and treatment related to women and children. The patients are treated free of cost. Patients are also given food, blanket, medicine, clothes etc., free of cost. This NGO also works for the relief of flood , epidemic, starvation, earthquakes, epidemic, storms victims. It distributes clothes, food, medicine, balanket, drinking water, etc. among victims as rlief measure. This NGO has its own staff, vehicles and offices.

4. *CARE India*

This NGO is related to an International NGO called CARE. Organisations of this NGO are working in different countries of the world. The headquarters of CARE India is located at Delhi, but this NGO has its organisations in all states and Union Territories of our country. This NGO has its own staff, vehicles and offices. This NGO is working for removal of malnutrition and diseases prevalent among women and children of weaker sections. Through this NGO ICDS Programmes are being run throughout the length and breadth of the country. This NGO also works for literacy, education, health, nutrition, sanitation and environment conservation. It also extends its services during flood, storm, cyclone, epidemic earthquake, starvation and other natural calamities.

5. Missionaries of Charity

This NGO was established by mother Teressa. This is also an International NGO working for the welfare and development of deprived and oppressed community all over the world. In India, its head office was at Kolkata. It has its organisation in different states of our country. Welfare works related to poor, disabled, handicapped, mentally retarded and Lepers are being done by this NGO.

6. Ramkrishna Mission

This is a national level NGO. Its headquarter is located at Narendrapur in Kolkata, West Bengal. Its branch offices are located throughout the length and breadth of the country. This NGO was started in 1896 by Late Swamy Vivekananda, the great social worker and reformer of India. Now it has become an international NGO because its offices have been established in foreign countries also. This NGO runs welfare and development works for STs and SCs in the field like education, agriculture, health, sanitation, training etc.

7. Arya Samaj

This NGO was established in 1875 by Swamy Dayananda Saraswati. The volunteers of this NGO are found in all parts of our country. They are extending their services for the welfare of deprived, exploited and oppressed communities.

8. Servants of India Society

This NGO was established at Poona in 1905 by Gopal Krishna Gokhale. This NGO is extending its services for the development and welfare of SCs, STs and other backward classes ever since its establishment.

9. Akhil Bhartiya Harijan Sewak Sangh

This NGO was founded by our Rastrapita Late Mahatma Gandhi in 1932. This NGO has been extending its services and cooperation in removing the practice of untouchability and eradicating the problem of illiteracy, health, nutrition sanitation, exploitation and oppression and social disability prevalent among the SCs of our country ever since its establishment.

10. All India Depressed Classes Federation

This NGO was founded by Late B.R. Ambedkar. This NGO has made valuable contribution in upliftment, welfare and awakening of SCs.

11. Thakkar Bapa Ashram

This NGO was established in Orissa at Nimkhandi by Late A.V. Thakkar, who is popularly known as Thakkar Bapa. It has fought for the cause of weaker sections even during British Raj.

12. Andhra Rastra Adim Jati Sewak Sangh

The headquarters of this NGO is located at Nellore in Andhra Pradesh. This NGO is contributing much in the development and welfare affairs of the weaker sections of the Andhra Pradesh state.

13. Social and Research Centre

This is an NGO whose headquarters is located at Tilona in Rajasthan. This NGO is working for alround development of weaker sections of Rajasthan state.

14. Satya Sodhak Samaj

This NGO was founded by Joytiba Phulle in 1837 in Maharashtra. Ever since its establishment, it is contributing efficiently towards welfare and developmnet of SCs and STs in the state of Maharashtra.

15. Hindu Sweepers Sevak Sangh

This is an NGO delicated and committed towards the welfare of Hindu sweepers. Its headquarter is located at Delhi. It has made valuable contribution towards the development of Hindu scavengers of Delhi.

16. Bhartiya Samaj Unnat Mandal

This is an NGO whose headquarters is located at Bhiwandi in Maharashtra. It has played significant role in the welfare and development of weaker sections.

17. Ishwar Sharan Ashram

The headquarter of this NGO is located at Allahabad. It has contributed a lot towards the welfare of weaker sections.

18. Vanvasi Kalyan Kendra

This is an NGO working for the welfare of the tribals. This NGO is working almost in all states having concentration of tribal population. Works related to schools, hospitals, health, sanitation, agriculture, training programmes, artifacts and handicrafts exhibition etc. are taken up by this NGO working in Tribal Areas.

NGOs Provided Grants-in-aid by the mininstry of Tribal Affairs in 1999-2000, 2000-2001 as on 15.2.2001

The mininstry of Tribal Affairs have provided Grants-in-aid to NGO during 1999-2000 and 2000-01 under the scheme of educational complex in Low Literacy pockets for Development of women literacy in Tribal Areas, scheme of vocational Training centres, Development of PTGs and promotion of voluntary efforts. A list of these NGOs are being provided for the sake of knowledge to the readers.

State-wise List of NGOs Provided Grants-in-aid during 1999-2000 and 2000-01 as on 15.2.2001 under the scheme of Educational Complex in Low Literacy Pockets in Tribal Areas

Sl. No.	*Name of Organisation*	*1999-2000*	*2000-2001*
	Andhra Pradesh		
1.	Little Flower Society, Medak (A.P.)	3,45,000	—
2.	Social Action for Social Development (SACD) Sanjeeva Reddy Nagar, Hyderabad	1,20,000	—
3.	Integrated Development Agency, Nandigram, Krishna	3,60,000	—
4.	Jagurithi Educational and Community Development Society	2,25,00	—
5.	Vennela Educational and Rural Development Society, Hyderabad	1,89,000	—
6.	Social Integrated and Rural Development Society, Panagal, Nalgonda	1,62,500	—
7.	Chatainya Rural Development Association Praddatur, Cuddapath	1,62,500	—
8.	SAM Society for Social Justice, Ranga Reddy, Andhra Pradesh	1,62,500	—
9.	Priyadarshani Service Organisation, Narshimha Nagar	1,62,500	—
10.	Chaitainya Educational and Rural Development Society, Cuddapah	1,62,500	—
11.	Rural and Urban Progressive society Chintal Basti, Hyderabad	1,62,500	—
12.	Sri Venkateshwara Mahila Mandali, Guntur, Andhra Pradesh	56,500	—
13.	Santosh Educational Society, Vidya Nagar, Karim Nagar	56,500	—
14.	Arundhati Educational Society, North Calaguda, Secundarabad	56,500	—
15.	Adarsh Gramin Girja Mahila Swachanada Seva Sansthan, Hamam Konda, Warangal	—	1,89,000
16.	Navodaya Integrated Cultural, Social, Educational and Voluntary Action, Karnool	—	1,89,000
17.	Guntur Rural Educational and Development Society, A.P.	—	1,89,000
18.	Rural Mahila Welfare Society, Kullur, Kurnool, A.P.	—	1,89,000

(contd.)

(contd.)

Sl. No.	*Name of Organisation*	*1999-2000*	*2000-2001*
Arunachal Pradesh			
19.	OJu Welfare Association, Pampunpara	3,35,000	—
Bihar			
20.	Rajendra Institute of Education and Social Welfare, West Champaran	2,02,500	—
21.	Vishwa Bandhu Ramavatar, Jamia Madhya Vidyalaya	2,61,000	—
Gujarat			
22.	Sri Manilal Ganga Das Patel Sarvodya Kendra, Palanpur	2,00,532	—
23.	Lok Niketan Ratanpur, Palampur	6,40,000	—
24.	Shri Sarvodya Ashram, Sanali, Banaskantha	2,25,000	64,50,000
25.	Gram Swaraj Sangh, Kutch	5,31,992	—
26.	Shrimali Shushilakan Memorial Trust	—	6,75,000
Madhya Pradesh including Chhittisgarh			
27.	Rural Development Service Society, Pushpa, Sopcial, Cube, Raisen	6,64,000	1,74,000
28.	Prasad Ekla Samiti, Arjunnagar, Raisen	4,56,000	—
29.	Gramin Seva Kendra, Mandalinathu	2,25,000	2,25,000
30.	Rajendra Ashram Trust, Kathiwad, Jhabua	4,59,000	—
31.	Bheel Seva Sangh, Jhabua	3,35,720	—
32.	Deendayal Research Institute	6,30,000	—
33.	Adarsh Shiksha Samiti, Village Bakia, Satna	8,19,000 56,000	—
34.	Sri Ramkrishna Vivekanand Sevashram, Sahdol	3,99,000	—
35.	Maha Kaushal Mahila Shiksha Samiti, Jabalpur Sahdol	1,02,000	—
36.	Kasturba Gandhi Memorial Trust, Kasturba Gram	79,636	—
37.	Vivekanand Institute of Social Health and Social Welfare	2,79,990	6,61,500
38.	Pranami Adivasi Shiksha Samiti, Sirsi Jagir, District Guna	—	2,42,700
39.	Bandhewal Shiksha Samiti, Bhopal	2,25,000	—
40.	Pramod Adivasi Vikas Parishad, Jabalpur,	1,89,5000	—
41.	Pushpa Convent Shiksha Samiti, Bhopal	1,62,500	—
42.	Asha Deep Kalyan Samiti, Shihora, Jabalpur	1,62,500	—
43.	Madhya Pradesh Adivasi Sevak Sangh, Amahia, Rewa	1,62,500	—
44.	Jagat Shiksha Awam Samaj Kalyan Samiti, Shivpuri	1,62,500	—
45.	Aradhna Gramin Seva Samiti, Mattan, Mohalla, Niwari	1,62,500	—

(contd.)

(contd.)

Sl. No.	Name of Organisation	1999-2000	2000-2001
46.	Shakti Swasthaya Awam Samaj Kalyan Samiti, Gwalior	1,62,500	—
47.	Shivam Social Development Society, Moorena	1,62.500	—
48.	Annapurna Shiksha Samiti, Samer Khappa (Achli)	52,100	—
49.	Richa Samaj Seva Awam Shiksha Prasar Samiti, Krishna Vihar, Gwalior	56,500	—
Maharashtra			
50.	Apang Association, Mayanagar, Amaravati at Village Madegaon, District Nasik	—	4,56,000
51.	Tantrik Prashikshan Sansthan, Pawannagar, Amarawati	3,36,000	—
52.	Jeevandhara Viyam Avam Krida Shikshan Prasarak Sansthan, Akola	57,500	—
53.	Sawitribai Jyotirao Phule Samaj Seva Sanstha, Akola	1,40,000	—
54.	Akhil Bhartiya Mgasvargiya Samaj Probhadan Sansthan, Thane	1,62,500	—
55.	Subha Lakshmi Shikshan Sansthan, Raigarh	—	1,77,00
Orissa			
56.	Marr-Munning Ashram, Jeypore, Koraput	3,62,500	3,37,500
57.	Nyay Sahayak Samiti, Koraput	3,90,00	3,60,000
58.	Janakalyan Samiti, Bhubaneshwar	3,90,000	—
59.	Bright Career Academy, Jeypore Koraput	—	7,50,000
60.	Bioparingnda Kestra Samiti Biopringnda, Koraput	135,982	3,37,500
61.	Liberation Education and Action for Development, Koraput	3,45,000	3,15,000
62.	Sarvodya Samiti, Koraput	3,49,500	—
63.	Seva Samaj, Rayagada	2,45,000	—
64.	Servants of India Society, Rayagada	4,50,000	—
65.	Agargamee, Kashipur, Rayagada	—	3,39,048
66.	Institute of Engineering and Management Jeypore	—	7,50,000
67.	Kasturba Gandhi Memorial Trust	4,55,500	—
68.	Society for Nature Education Health, Bhubaneshwar	—	9,14,000
69.	National Youth Service Action and Social Development Research Institute, Dhenkanal	8,39,000	—
70.	Social Education for Environment and Development, Bhubaneshwar	1,95,000	—

(contd.)

(contd.)

Sl. No.	*Name of Organisation*	*1999-2000*	*2000-2001*
71.	Arun Institute of Rural Afairs, Kurnool, Dhenkanal	56,500	—
72.	Sri Ramkrishna Ashram, Angul	56,500	—
73.	Parkalpa, Keonjhar	—	2,15,000
	Rajasthan		
74.	Rajasthan Balkalyan Samiti, Jodhol, Udaipur	2,25,000	—
75.	Mahavir Jain Vidyalaya, Sansthan, Udaipur	54,000	—
76.	Mewar Sharirik Shiksha Samiti, Udaipur	4,59,000	—
77.	Om Vidya Mandir, Tonk	—	2,79,000
78.	St. Soldier Education Society, Tonk	4,59,000	—
79.	Abhinav Shiksha Samiti, Nav Aabadi, Banaswara	2,67,000	—
80.	Zila Yuva Meena Samaj Sansthan, Bhilwara	70,470	—
81.	Janajati Mahila Vikas Sansthan, Sawai Madhopur	—	1,65,000
	Uttar Pradesh Including Uttarakhand		
82.	Hindu Muslim Ekata Kalyan Samiti, Lal Kuan, Luchnow	—	4,56,000
83.	Gramin Mahila Silai, Kadai and Bunai Kendra, Azamgarh	1,26,000	3,30,000
84.	Dr. Ambedkar Shiksha Sansthan, Maharaj Ganj, U.P.	1,62,500	—
85.	Bhawani Gramodyog Seva Sansthan, Maharaj Ganj, U.P.	1,62,500	—
86.	Bhartiya Samajothan Seva Sansthan, Deoria, U.P.	1,62,500	—
87.	Manav Vikas Awam Seva Sansthan, Lucknow	—	1,89,000
88.	Ambedkar Shiksha Samiti, Lucknow	—	4,59,000
	West Bengal		
89.	Bharat Sevashram Sangh, PO Beldanga, Murshidabad	—	2,78,000

List of NGOs Provided Grants-in-aid during 1999-2000, 2000-01 Under Vocational Training Centre Scheme

Sl. No.	*Name of Organisation*	*1999-2000*	*2000-2001*
Andhra Pradesh			
1.	Integrated Development Agency, Nandigram, Krishna	3,60,000	—
2.	Priyadarshani Service Society, Padru, Vishakhapatnam	—	9,63,000
3.	Women Awareness and General Education Society, Vijyavada	3,60,500	—
4.	A.P. Girijan Sevak Sangh, Nandigram, Krishna	3,60,500	—
Madhya Pradesh Including Chhattisgarh.			
5.	Bandhewal Shiksha Samiti,Village Bilikesh Ganj, Sehore	3,60,500	—
6.	Gramin Swasthya Awam Samaj Kalyan Sansthan, Shivpuri	8,42,500	—
7.	S.P. Memorial Shiksha Niketan Samiti Jabalpur	3,60,000	—
8.	Satyavati Savitri Shiksha Awam Samaj Kalyan Samiti, Gwalior	8,42,500	—
9.	Amar Mahila and Bal Uthan Samiti Village Saranli, Majhgaon, Block Sihora, Jabalpur	—	9,63,000
Maharashtra			
10.	Dr. Balasaheb Ambedkar Adivasi Magaswrajiya Shiksha Sansthan, Amravati	—	9,63,000
11.	Kasturba Gramin Mahila Shikshan Prasarak Mandal, Akola	—	9,63,000
12.	Social Welfare Society, Village Siswa Barkar, P.O. Silhari Sardkho, Basti	3,60,500	—
Assam			
13.	Dr. Ambedkar Mission, Village Dhopatri, Kamroop	3,60,500	—
Nagaland			
14.	Mulharukhel Welfare Society Dzelhamiphek	3,60,500	—
15.	Keize Society, Lerie, Kohima	—	9,63,000
16.	Wokha Town Women Welfare Society	—	9,63,000
17.	Women Welfare Society, lunchebato	—	9,63,000
18.	Vitok Women Society, New Market, Kohima	—	9,63,000
West Bengal			
19.	Gandhi Gram Sarak Seva Kendra Midnapore	7,23,000	—
20.	West Bengal SCs and STs and Minorities Welfare Association, Kolkata	7,23,000	—
21.	All India Federation of SCs and STs and Minorities, Kolkata	7,23,000	—
22.	Satdyju Mahila Mandal Midnapore	7,23,000	6,02,500

List of NGOs Provided Grants-in-aid for the Development of Primitive Tribal Groups (PTGs)

Sl. No.	*Name of Organisation*	*Amount Released in 1999-2000 As on 15.2.2001*
1.	Swami Vivekanand Youth Movement, Mysore, Karnataka	4,02,010
2.	Integrated Rural Development Welfare Association, Imphal, Manipur	9,73,800
3.	Kalinga Institute of Industrial Technology Bhubneshwar, Orissa	19,02,905
4.	Prakalpa, Keonjhar, Orissa	6,50,905
5.	Society for Welfare of Weaker Sections, Gajapati, Orissa	3,38,000
6.	Madhya Pradesh Vigyan Sabha, Bhopal, MP	3,88,000
7.	Dharampur Utthan Vahini, Navsari, Gujarat	14,05,000
8.	Shakti Pratisthan, Ahmedabad, Gujarat	2,75,000
9.	Nehru Yuva Kendra, Malanpuram, Kerala	25,63,000
10.	A.P.K. Science and Social Organisation, Agartala, Tripura	8,87,000

Statewise List of NGOs. Supported by the Ministry of Tribal Affairs, Government of India

Sl. No.	*Name of Organisation*	*Project*	*Amount Released 1999-2000*	*Amount Released 2000-2001 As on 15.2.2001*
Andhra Pradesh				
1.	Christ Rural Development and Education Society, Anantapur	Residential School	9,83,881	—
2.	Centre for Rural Education and Development Society, Gorantla, Anantpur	Residential School	9,96,120	—
3.	Gram Abhiuday Society for Integrated Development, Urvakonda, Anantpur	Residential School	3,85,839 8,24,553	—
4.	Human Resource Development Association Punganur, Chittoor	Residential School	5,86,710	4,74,220 10,60,920

(contd.)

(contd.)

Sl. No.	*Name of Organisation*	*Project*	*Amount Released 1999-2000*	*2000-2001 As on 15.2.2001*
5.	Community Health and Rural Development, Palamaner, Chittoor	5-Balawadis Tying and Shorthand	2,01,397 2,01,398	
6.	Praja Abhiudaya Seva Samiti Greanspet, Chittoor	Computer and Residential School	12,42,900	
7.	Rural Organisation for Social Activity, Mandela Guntur	Residential School	575,910 4,02,921	
8.	Rural Educational and Development Society, Guntur	Mobile Dispensary	3,07,932 1,95,431	—
9.	Society Action for Social Development, Hyderabad	15-Creches	3,37,387 3,37,388	—
10.	A.P. Girijana Sevak Sangh, Nandigama, Krishna District	Residential School and 10-Bed Hospital	10,10,160	9,02,040 724,023
11.	Vani Niketan Mahila Mandali, Nandigama, Krishna District	Non-residential School	2,08,701	—
12.	Integrated Development Agency, Nandigama, Krishna District	Mobile Dispensary and 10-Bed Hospital	11,48,610	—
13.	Ram Krishna Mission Rajamundari, AP	2 Mobile Dispensaries	1,11,181	214,335 2,14,335
14.	Social Transformation and Rural Technology, Kukatapally, Hyderabad	Residential School	5,35,220 5,25,610	— —
15.	Swan Education Society Rangareddy	Mobile Library Hostel Non-Residential School Typing and Shorthand	6,08,659 7,05,825	— —
16.	R.K. Mission Vishakhapattanam	2-Mobile Dispensary	4,14,360 4,14,360	3,68,296 4,28,670
17.	Telangana Weaker Section Society, Adilabad	MobileDispensary	4,89,008	—

(contd.)

(Contd.)

Sl. No.	*Name of Organisation*	*Project*	*Amount Released 1999-2000*	*2000-2001 As on 15.2.2001*
18.	Bharath Integrated Rural Development Society, Ananthapur	Non-Residential School	—	3,99,465
19.	Dr Ambedkar Dalitha-wanga Abhiruddi Sangam, Chuddapah	Residential School	—	7,59,510
20.	Sri Venketeswara Mahila Mandali, Guntur	10-Bedded Hospital	—	12,90,728
21.	People's Educational Society, Karim Nagar	Residential School (P)	—	7,48,710
22.	Society for Education and Rural Technology, Karim Nagar	Residential School	—	7,59,110
23.	Sri Sairam Seva Sangam, Karimnagar	Residential School	—	7,10,910
24.	Vijaya Educational Society, Karimnagar	Residential School	—	7,59,510
25.	Bapuji Integrated Rural Developmemt Society, Kondru, Krishna	Residential School	—	7,59,510
26.	Raghavendra Rural Development Society, Kurnool	Computer Training Centre	—	6,59,340
27.	St.Ann's Educational Society, Kurnool	Non-Residential School	—	4,37,085
28.	Chetanya Educational Society, Kurnool	Computer Training Centre	—	6,59,610
29.	Society for Humanity, Kurnool	Rural Night School	—	1,42,778
30.	Health Agriculture Rural Development Society, Kurnool	Non-Residential School	—	3,99,465
31.	Bharat Ratna Mother Teresa Educational Academy, Kurnool	Residential Workers Training Centre	—	3,46,605
32.	Constructive Workers and Development Society, Mahboob Nagar	Residential School	—	4,85,190

(Contd.)

(Contd.)

Sl. No.	Name of Organisation	Project	Amount Released 1999-2000	2000-2001 As on 15.2.2001
33.	SURPASS, Medak	Mobile Dispensary 10-beded Hospital	—	17,70,736
34.	Sri Mandalapu Narayana Education Society, Rangareddy	Residential School	—	7,59,510
35.	Youth in Action Society, Rangareddy	10-beded Hospital	—	12,90,728
36.	Sri Sai Educational Society, Rangareddy	Non-residential School	—	3,99,665
37.	Priyadarshani Service Organisation, Vishakhapattanam	Residential School	—	7,59,510
38.	Sri Parmeshware Educational Society Atamkur, Kurnool	Mobile Dispensary	—	4,24,418
39.	Sri Govindaraja Swamy Social Development Society, Cuddapah	Residential School	—	4,20,866
40.	Gautam Educational Society, Secundarabad	Non-Residential School	—	1,66,424
41.	Baba's Organisation for Social Services, MIG-19 Warranagal District	Residential School	—	5,95,643
42.	Grameena Seva Society, Kurnool	Workers Training Centre, Residential	—	3,04,605
43.	Social Action in Development	Non-residential School	—	3,99,665
Arunachal Pradesh				
44.	R.K. Mission Hospital Itanagar	60-beded Hospital	21,84,705 21,84,705	21,66,705 21,66,705
45.	Arunachal Pali Vidyapeeth, Lohit	Residential School	5,61,060	4,48,560
46.	OJU Welfare Association, Naharlagun	Residential School Hostel	7,55,955	
		Mobile Dispensary	17,69,039	2,58,660

(contd.)

(Contd.)

Sl. No.	*Name of Organisation*	*Project*	*Amount Released 1999-2000*	*2000-2001 As on 15.2.2001*
47.	Dony Polio Mission, Papumpare	3 Hostels 10-Beded Hospital	29,34,270	21,47,670
48.	Centre for Buddhist Cultural Study, Tawang	Residential School	10,54,575	6,12,360
49.	R.K. Mission, Narottam Nagar (Tirap)	Computer-2 units Mobile Dispensary	3,09,200 3,09,200	3,98,580 5,60,620
50.	Buddhist Cultural Preservation Society, Bomdilla, W. Kamong	Residential School	—	19,7,190
51.	R.K. Mission (Along) West Slang	Non-Residential School Hostel	31,65,275	—
		School Bus (3) Mobile Medical A.V.unit 10-Beded Hospital	29,81,525	34,83,170
52.	R.K. Mission School, Narottam Nagar, Tirap	5-Residential Schools	21,58,495 13,85,154	28,89,701 11,03,329
Bihar and Jharkhand				
53.	Bharat Sevashram Sangh (Sonari) Singhbhum (East and West)	3-Residential School Mobile Dispensary	17,76,902	—
		10-beded Hospital with Mobile Dispensary and Arrear grant	14,62,690	24,67,783 22,68,662 6,45,977 3,79,755 1,98,230
54.	Bharat Sevashram Sangh (Ghatsila) East Singhbhum)	Mobile Dispensary	1,97,760	—
		10-beded Mobile Dispensary Residential School Training in Carpentry	2,84,760	26,36,325
55.	Alpasankhyak Mahila Prashikshan Sansthan, Godda	Mobile Dispensary	3,93,316	—
56.	R.K. Mission, Viveka Nanda Society, Bistupur, Jamshedpur	Hostel, Library cum Audio Type	2,52,095	
		cum Shorthand Computer	2,56,023	4,09,556

(Contd.)

(Contd.)

Sl. No.	Name of Organisation	Project	Amount Released 1999-2000	2000-2001 As on 15.2.2001
57.	Zila Adivasi Naitik Utthan Sangh, Purnia	Non-Residential School	1,90,926	—
		Divyayan Unit (Agriculture)	47,59,931	—
58.	R.K. Mission Ashram Morabadi, Ranchi	Rural Night School	27,82,960	25,28,184
		Mobile Dispensary		3,26,193
		Mobile Library		
		A.V. Unit		
59.	R.K. Mission Tuberculosis Sanatorium, Ranchi	Hospital (80-beded)	12,21,525	—
		Mobile Dispensary	31,32,045	7,52,109
60.	Janajati Vikas Pratisthan, Topchachi, Dhanbad	10-beded Hospital	—	12,90,728
Himachal Pradesh				
61.	Rinchen Zangpo Society for	Spiti Residential School	8,50,000	—
	Development, Dharamsala		6,54,000	8,28,114
62.	Himalayan Buddhist Cultural Assoiation, Manali, Kullu.	Residential School	1,13,301	16,57,539
63.	Ramdha Society, Rangrik, L & S District	Hostel	1,26,178	4,70,583
64.	The Institute of Studies in Buddhist Philosophy and Tribal Culture Society, Tabo	Residential School	—	3,98,366
Gujarat				
65.	Bharat Sevashram Sangh, Ahmedabad	Mobile Dispensary	—	5,66,865
66.	Dhadhela Kelawani, Mandal, District, Dahod	Hostel	2,42,730	2,74,023
67.	Bharat Yatra Kendra, Narmada	Hostel, Mobile Dispensary Typing and Shorthand	8,23,621	—
68.	Larpan Nasapur Vibhag, Kelavani Mandalam Surat	Girls Hostel Boys Hostel	6,70,995	—

(Contd.)

(Contd.)

Sl. No.	*Name of Organisation*	*Project*	*Amount Released 1999-2000*	*2000-2001 As on 15.2.2001*
69.	International Rural Education and Cultural Association, Bharuch	Hostel	5,39,730	—
70.	Bharat Sevashram Sangh, Nawasari	Mobile Dispensary Mobile library-cum-A.V. Unit	14,10,528	—
71.	Bharat Research and Publication Centre, Vadodara	Library	51,675	—
	Jammu and Kashmir			
72.	All India Centre for Urban and Rural Development, New Delhi	Typing, Shorthand and Computer	7,68,960	5,19,480
73.	Buddhist Youth Association, Kargil	4-Residential School Dispensary	1r,99,985	—
74.	Lamdon Society Welfare Society, Ladakh	Residential School	14,45,987	—
75.	Lamdan Social Welfare Associal, Kargil	Non-Residential School (New)	—	6,69,330
76.	Buddhist Youth Association, Kargil	Non-Residential School-3	12,53,745	
77.	Himalayan Buddhist Cultural Society, Doda	Non-Residential School	—	1,42,933
78.	Gujar Desh, Charitable Trust, Bye Pass, Jammu-Tawi	Residential School	—	4,49,218
	Kerala			
79.	Sri R.K. Avaita Ashram Ernakulum	Hostel	5,03,415	5,05,665
80.	Swamy Nirmla Nand Memorial Mission, Balbhagwan, Kayamkulam	Hostel	2,13,930 3,07,505	2,50,200 1,98,450
81.	Swamy Vivekananda Memorial Mission, Muttil Waynad	Mobile Dispensary 10-beded Hospital	2,00,835 10,04,625	8,30,001 1,91,123
82.	Amrita Bhawanam, Mata Amrita Nanda, Mayidevi Math, Pripall	Hostel	—	17,14,560

(Contd.)

(Contd.)

Sl. No.	Name of Organisation	Project	Amount Released 1999-2000	2000-2001 As on 15.2.2001
83.	Harijan Sevak Sangh, Kerala Branch Thiruvananthapuram	Typing 6 Chreches, Computer	— —	3,99,240 6,63,308
84.	Vinoba Niketan Thiruvananthapuram	Hostel Mobile Dispensary	— —	9,76,080 5,23,905
85.	Vanvashi Ashram Trust, Waynad	Residential School	3,11,010	18,25,343
New Delhi				
86.	Akhil Bhartiya Dayanand Sevak Sangh, New Delhi	Hostel (Nagaland)	3,83,385 7,48,980	10,60,627 4,30,920
87.	Bhartiya Adimjati Sevak Sangh, New Delhi	Hostel, Tribal Workers Training Tribal Women Workers 4-tribal Kanya Ashram	36,06,147	20,40,135
88.	Deendayal Research Institute, New Delhi	Hostel-2	6,04,800	
Orissa				
89.	Bhairavi Club, Khurda	Residential	5,11,560	—
90.	Patriot, Bhubaneshwar	Mobile Dispensary	3,90,764	1,89,910
91.	R.K. Mission, Bhubaneshwar	Hostel Library	2,51,982 2,51,982	3,22,137 3,53,646
92.	Orissa Socio-economic Development Association, Bhubaneshwar	Computer	2,16,360	—
93.	Kalinga Institute of Industrial Techlology, Bhubaneshwar	Residential School	—	5,87,920 16,62,673
94.	Research and Analysis Consultants, Bhubaneshwar	Computer	3,52,440,	—
95.	Animal Welfare Society of Orissa	Mobile Dispensary	—	1,95,435
96.	Orissa Social Rural Technology Institute, Cuttak	Mobile Dispensary	— —	1,95,435 2,81,785

(Contd.)

(Contd.)

Sl. No.	*Name of Organisation*	*Project*	*Amount Released*	
			1999-2000	*2000-2001 As on 15.2.2001*
97.	Banki Achalika Harijan Adivasi Welfare Parishad, Cuttak	Hostel 5-Creches	— —	4,87,418 4,09,792
98.	Cuttak Zila Harijan Adivasi Seva Sanskar Yojna, Cuttak	Residential, School (Primary)		57,591 4,63,410
99.	Community Legal and Research Centre, Dhenkanal	Non-Residential School	—	1,66,403
100.	Arun Institute for Rural Affairs, Dhenkanal	Residential School (P)	5,40,810 2,28,310	 5,92,110
101.	Laxmi Narayan Harijan Adivasi Backward Development Society, Jaipur	Residential School	10,39,320	—
102	Natural Institute for Social Welfare, Jaipur	Typing Shorthand	1.31,670	—
103.	Shree Ramkrishna Ashram, Kalahandi	Student Home Mobile Dispensary	4,20,460 3,69,110	3,68,910 4,20,660
104.	Ramkrishna Vedanta Ashram Kalahandi	Mobile Dispensary	3,27,120	—
105.	Vanvasi Sewa Samiti, Kandhamal	Hostel	4,19,130 3,15,630	—
106.	Vishwa Jeevan Seva Sangh, Khurda	Residential School (P)	5,19,660 5,19,660	—
107.	Nehru Seva Sangh, Khurda	Hostel	5,92,380 4,37,130	
108.	Organisation for Rural Women and Youth Development, Mayurbhanj	Residential School	8,83,620	5,81,210
109.	Harijan Adivasi Kanya Sevashram, Nayagarh	Non-Residential School, Mobile Dispensary	13,75,830	—
110.	Social Weaker Awareness Development and Economic Service Institute . (Swadesi Phulbane)	Residential School (P)	51,19,660 51,19,660	

(Contd.)

(Contd.)

Sl. No.	*Name of Organisation*	*Project*	*Amount Released 1999-2000*	*2000-2001 As on 15.2.2001*
111.	R.K. Mission Ashram, Puri	Mobile Dispensary Hostel	10,36,273 2,65,547	5,29,802 4,57,350
112.	Seva Samaj, Rayagada	Typing Training, Hostel	4,45,410 1,22,580	—
113.	Orissa Sarovaya Parishad, Rourkela	2-Hostel	5,02,220	4,50,450
114.	Samaj Kalyan Sansthan, Mayurbhanj	Balwadi Centre and Residential School		5,19,660
115.	Association for Voluntary Action, Puri	Residential School	5,49,360	—
116.	Jeevan Jyoti, Bhubaneshwar	Ten beded Hospital	12,27,278	—
117.	Manav Sevasadan, Deogarh	Non-Residential School	—	3,92,040
118.	Vidhut Club, Puri	Non-Residential School	—	4,07,790
Rajasthan				
119.	Bansthali, Vidyapeeth, Bansthali, Rajasthan	Hostel	24,48,700	10,61,215
West Bengal				
120.	Birsamunda Education Centre, Jalpaiguri	Residential School	5,66,370	—
121.	Society for Rural and Urban Development. Paragamas Harijan Sevak Sangh, Howrah	Residential School (S) 22 Creches/ Balwadi	3,64,558 4,98,934	— —
122.	Bharat Sevashram Sangh, Birbhum	Hostel	1,53,203 1,23,367	—
123.	Gandhi Sarak Gram Sevak Kendra, Kolkata	Residential School	5,86,710 —	 4,74,210
124.	All Inda Federation of STs/SCs and Minorities, Kolkata	Residential School	5,87,710	— 4,74,220
125.	Institute of Social Research and Applied Anthropology, Midnapore	Balwadis-5		58,590

(Contd.)

(Contd.)

Sl. No.	Name of Organisation	Project	Amount Released 1999-2000	2000-2001 As on 15.2.2001
126.	Bharat Sevashram Sangh, Dokra, Midnapore	Residential School Mobile Dispensary	13,58,633	7,35,570
127.	Bharat Sevashram Sangh (Aurangabad), Murshidabad	Mobile Dispensary	7,25,850	5,63,985 3,11,310
128.	Bharat Sevashram Sangh (Beldenga), Murshidabad	Hostel	2,05,830	7,25,310 4,57,920
129.	West Bengal SCs/STs and Minority Welfare Association, Midnapore	10-beded Hospital,Mobile Library-cum-AV.	6,05,115	6,05,115
130.	R.K. Mission, Boy Home, North Parganas	Hostel	2,24,190 2,24,190	2,78,125 3,93,600
131.	Bharat Sevashram Sangh, Dakshin Dinajpur	Hostel	—	2,28,930
132.	Bharat Sevashram Sangh, Balurghat, South Dinajpur	Hostel	1,13,160	6,38,303 5,01,277
133.	Bharat Sevashram Sangh, Hoogly	Hostel Library		6,96,040
134.	Dhaniakhali Depressed Classes League, Hoogly	Residential School	—	7,48,710
135.	Gohaldina Jati, Upjati Blue Bird Welfare Centre, Midnapore	Residential School(P)		6,47,884
136.	Bharat Sevashram Sangh, Beldanga, Murshidabad	Mobile Dispensary	—	6,44,497
137.	Bharat Sevashram Sangh, Nadia	Residential School and Mobile Dispensary	—	15,32,517
138.	Bharat Sevashram Sangh, Aurangabad	Mobile Dispensary	—	5,83,309
139.	Vivekananda Child Welfare Home, Kakdwip	Residential School	5,75,910	—

(Contd.)

(Contd.)

Sl. No.	Name of Organisation	Project	Amount Released 1999-2000	Amount Released 2000-2001 As on 15.2.2001
140.	Himalayan Buddhist Cultural Association, Jalpaiguri	Non-Residential School	—	3,88,700
141.	Nikhil Bharat Vanvashi Panchayat, Midnapore	Residential School 16-Balwadis	—	2,19,240
Madhya Pradesh and Chhattisgarh				
142.	R.K. Mission Ashram, Narainpur, Bastar	Hotel and Automobile Engineering	4,70,565 4,70,565	5,51,627 5,51,627
143.	Rama Education and Welfare Society, Bhopal	Non-Residential School Computer Centre	7,51,239	—
144.	Bandhewal Shiksha Samiti, Bhopal	Computer Non-Residential School	7,51,463	—
145.	Asha Deep Kalyan Samiti, Jabalpur	Non-Residential School Craft Centre, Computer Training 10-Bed Hospital	19,04,611	—
146.	Adivasi Vikas Samiti, Mandala	Non-Residential School	3,25,170	—
147.	Gram Chetna Seva Samiti, Morena	Residential School	1,67,303	—
148.	Shri Shanti Niketan Shiksha Prachar Samiti, Morena	Computer	5,84,460	—
149.	Lata Jaganath Shiksha Prasar Awam Samaj Kalyan Samiti, Sagar	Non-Residential School	1,68,203	—
150.	Lovely Samaj Seva Evam Shiksha Prasar, Samiti, Sagar	Non-Residential School	1,68,203	—
151.	Shri Lal Bahadur Shiksha Prasar Samiti, Sagar	Non-Residential School	1,68,203	—
152.	Late Pyarelal Shiksha Samiti, Sagar	Non-Residential School	1,68,203	—

(Contd.)

(Contd.)

Sl. No.	*Name of Organisation*	*Project*	*Amount Released 1999-2000*	*2000-2001 As on 15.2.2001*
153.	M.P. Anusuchit Jati, Janajati Pichhada Varg, Ujjain	Residential School(P)	3,01,203	7,21,080 5,86,710
Assam				
154.	Garo Welfare and Rehabilitation Society, Guwahati	Computer Residential School Non-Residential School Library-cum Audio Visual unit 10-bed Hospital, Mobile Dispensary	 8,95,918 5,28,881	 3,57,168 17,70,736
155.	R.K. Mission Ashram, Ullubari, Guwahati	Mobile Dispensary Hostel Library	3,70,122 4,22,236	4,67,586 —
156.	Dr. Ambedkar Mission Kamrup	Mobile Dispensary 10-beded Hospital	8,76,135	—
157.	Prantiya Samaj Kalyan Kendra, Lakhimpur	Weaving Training Hostel Knitting Mobile Dispensary	—	3,42.972
158.	R.K. Mission, Sevashram, Sichar	Student Home	7,60,495	1,90,175 1,90,175
159.	Sree Manta Sankar Mission, Anglong	Mobile Dispensary	—	4,80,008
160.	Shishu Shiksha Samiti Guwahati	Non-Residential School	—	4,23,090
161.	Sadau Awam Gramya Puthi Bharat Sansthan, Nagaon	Library	—	85,050
162.	Pathari Vocational Institute, Nagaon	Computer	—	6,58,260 61,109
163.	Purvottar Janajati Shiksha Samiti, Tezpur	Residential School	—	7,44,210
Karnataka				
164.	Deorajurs Educational Trust, Bidar	Residential School	4,27,410	—
165.	Vivekananda Girijan Kalyan Kendra, Chamrajangar	Residential School Mobile Dispensary	5,73,322 8,14,109	

(Contd.)

(Contd.)

Sl. No.	Name of Organisation	Project	Amount Released 1999-2000	Amount Released 2000-2001 As on 15.2.2001
166.	Harihara Gramin Abhirudhi Sangh Kollar	10-beded Hospital Mobile Dispensary	9,11,692 3,04,124 1,95,435	1,33,530
167.	Swamy Vivekananda Youth Movement, Mysore	Residential School Mobile dispensary 10-beded Hospital	15,63,567 13,35,667	24,63,362
168.	Swamy Sarvandharama Shranalaya, Tumkur	10-beded Hospital, Non-Residential School	9,56,577	—
169.	Shri Hoysala Vidyasamathi, Davangere	Residential School Mobile Dispensary	—	12,39,518
170.	Bharathi Educational Trust, Kolar	Residential School	—	7,20,180
171.	Kumuwathi MA Rural Development Society, Tumkur	Mobile Dispensary	—	4,80,008
172.	Manjunatha Swamy Vidya Sansthan, Devangere	Residential School	—	6,75,746
Maharashtra				
173.	Sevadharmo Trust, Pune	Mobile Medical Unit	2,59,009	—
174.	ABM Samaj Prabodhan, Thane	Balwadi-50	—	11,61,000
175.	Chidam Adivasi Seva Mandal,Wardha	Hostel	—	9,85,005
176.	Dharmasamanavay Mah. Workers Vikas Shikshan Sansthan, Amravati	10-beded Hospital	—	5,35,455
177.	Shri Hanuman Vyayam Prasarak Mandal, Amravati	Computer Training Centre	—	3,19,679
178.	Ashtvinayak Bahuuddisheya Kalyankari Seva Samiti, Warda	Residential School	—	4,76,273
179.	The Family Planning and Medical Aid Trust, Mumbai	Mobile Library-cum-Audio-Visual Unit	—	12,65,354

(Contd.)

(Contd.)

Sl. No.	Name of Organisation	Project	Amount Released 1999-2000	2000-2001 As on 15.2.2001
	Manipur			
180.	Resource Centre for Social Welfare Community Development, Chandel	Residential School	3,08,757	—
181.	United Rural Development Service, Chandel			
182.	Sumchinvam Women Society, Churachandpur	Mobile Library-cum-A.V. Mobile Dispensary	—	1,33,165
183.	Moulvajphei Rural Health and Research Centre, Churchandpur	Mobile Dispensary	—	5,13,270
184.	Sumchinvam Youth Welfare Association, Churchandpur	Residential School	—	2,38,005
185.	Siamsinpawlpi Paite Student Welfare Assciation, Churchandpur	Residential School	14,76,568	14,76,567
186.	Dr. Ambedkar School, Planning and Development Society, Churchandpur	Residential School	—	11,52,720
187.	Institute of Social Development for Weaker Section, Imphal	Hostel	—	1,74,240
188.	Chil Chil Asian Mission Society, Imphal	Hostel	—	7,13,160
189.	Annallon Christian Development Society, Imphal	Residential Hostel	—	4,98,060
190.	North Eastern Boys Scout and Guide Association, Imphal	Residential School	—	11,86,920
191.	Onnuri Mission Society, Kanglatongbe	Residential School	—	7,82,820

(Contd.)

(Contd.)

Sl. No.	Name of Organisation	Project	Amount Released 1999-2000	Amount Released 2000-2001 As on 15.2.2001
192.	Manipur Eastern Hill Peoples Development Society, Lamphel	Residential School (P) Computer Training Typing Training	3,94,960	11,08,890
193.	Don Bosco Social Welfare Centre, Tamenglong	Hostel	—	9,61,340
194.	Centre of Rural Upliftment Service, Thoubal	Typing and Shorthand	—	1,66,320
195.	Integrated Rural Development and Educational Organisation, Thoubal	Residential School(P)	2,13,480	5,19,660 7,60,845
196.	Type Writing Institute and Rural Development Services, Thoubal	Residential School	—	8,88,372
197.	Rural Education Socio-Economic Development, Bishnupur	Non-Residential School	—	3,98,565
198.	Integrated Rural Development and Education Organisation, Chandel	Residential School	7,60,845	
199.	Manipur Tribal Farmers Development Association, Imphal	Computer Training	6,58,260	
200.	Ukhrul Distt. Multi-purpose Development Organisation, Imphal	Computer Training	—	6,55,560
201.	Rural Development Association, Imphal	Computer Training	—	6,55,560
202.	Trinity Institute for Rural Development, Tamenglong	Hostel	—	5,74,830
203.	Society for Women and Education Action and Reflection, Thoubal	Mobile Dispensary	—	4,85,408

(Contd.)

(Contd.)

Sl. No.	Name of Organisation	Project	Amount Released 1999-2000	2000-2001 As on 15.2.2001
204.	Integrated Educational Social Development Organisation	Residential School	5,81,760	—
205.	Wengling Tentha Farmers Development Association, Thoubal	Mobile Dispensary	—	4,85,407
206.	Volunteers for Rural Health and Action, Manipur	Mobile Dispensary	—	4,85,408
	Mizoram			
207.	HHAM Chantu, Pawl, Aizawal	Cane and Bamboo Crafts	89,820	—
208.	Mizoram Hmethai (Widows) Organisation, Aizawl	Mobile Dispensary, Residential School (S)	—	6,32,610
209.	Social Guidance Agency, Aizawl	Mobile Midical Unit	1,86,939	
210.	Rural Development Society, Chtimtupul	Residetial School	5,45,903	—
211.	Centre for Rural Development and Research	Mobile Dispensary	—	4,35,097
212.	Mizoram Upa Pawl, Aizawal	Mobile Dispensary	—	4,85,407
213.	Sacred Heart Society, Aizawl	Residential School	—	8,57,145
214.	Thewutak Numputtu Team, Aizawl	Carpentry Cane and Bamboo Craft Training Residential School	5,75,100	1,53,450
215.	Orental Institute Trust, Aizawl	Hostel	4,19,130	
	Nagaland			
216.	Nagano Society, Dimapur	Girls Hostel	2,96,733	—
217.	Nagaland Children's Home, Dimapur	Hostel	8,08,592	—

(Contd.)

(Contd.)

Sl. No.	Name of Organisation	Project	Amount Released 1999-2000	Amount Released 2000-2001 As on 15.2.2001
218.	Insukmok Kong Women Welfare Society Mokokochung	Self-Employment to Weaver	1,24,214	—
219.	Mecong Pang, Mepuweavers Association, Mokokchung	Weaving Training Centre	1,55,714	
220.	Aonglenden Weavers Association Mokokchung	Handloom and Weaving	1,55,714	
221.	Hill Welfare Society, Mokokchung	Self-Employment to Weavers	1,22,864	—
222.	Longlen Women's Society, Tusensang	Handicraft Training Centre	33,459	—
223.	LHIWEU Yetho Society, Kohima	Computer Training Centre	6,20,910	—
224.	Instegrated Rural Development Centre, Dimapur	Computer Training Centre	—	6,58,260
225.	Shan Woman Society, Dimapur	Non-Residential School	—	4,07,565
226.	Chewang Society, Kohima	Non-Residential School	—	5,11,065
227.	Nisaalem Economic Women Society, Kohima	Computer Training, Type and Shorthand	—	8,44,020
228.	AVI Society, Kohima	Computer Traning and Type and Short Hand	—	8,49,420
229.	Tong Pang Mepu Weavers Association, Mokokchung	Weaving and Handloom Centre		2,57,490
230.	Mongsen Mepu Weavers Association, Mokokchung	Handloom and Weaving Training Centre	—	2,57,490
231.	Mela and Yimsen Weavers Association, Mokokchung	Weaving and Handloom Centre	—	2,57,490

(Contd.)

(Contd.)

Sl. No.	*Name of Organisation*	*Project*	*Amount Released 1999-2000*	*2000-2001 As on 15.2.2001*
232.	SAARDAM Multipurpose Cooperative Society, Phek	Hostel	—	3,03,480
233.	Chizami B.S Multipurpose Society, Phek	Hostel	—	5,66,730
234.	Tuensang Women Welfare Society, Tuensang	Residential School	—	7,50,045
235.	Backward Development Society, Tuensang	10-bed Hospital	—	12,57,705
236.	Tizit Women Welfare Society, Nagaland	Knitting and Weaving Centre	—	1,72,830
237.	Mangkoterang Mepu Weavers Association, Mokochung	Handloom and Weaving Training Centre	—	2,57,490
238.	Luetang Kong Mepu Weavers Association, Mokokchung	Weaving and Handloom Training Centre	—	2,57,490
Tamil Nadu				
239.	Karuna Centre for Social Action, Anna District	25 Creches	1,19,362	—
240.	Sri C.P. Rama Swamy Aiyer Foundation, Chennai	Pottery and Graphic	2,38,500	—
241.	Tamil Nadu Village Consumer Protection, Council, Thiruvallore	Typing and Shorthand	1,25,234	—
Tripura				
242.	R.K. Mission, Agartala	Computer Unit Hostel	99,913	6,30,198
		Mint Dams (3) Mobile Dispensary	—	1,17,771
Uttar Pradesh				
243.	Gramin Mahila Silai, Kadai, Bunai Kendra, Alamgarh	Residential School (P)	5,65,713	—
244.	Samkon Viklang Kalyan Samiti, Chitrakut	Non-Residential School	6,04,875	—

(Contd.)

(Contd.)

Sl. No.	*Name of Organisation*	*Project*	*Amount Released 1999-2000*	*2000-2001 As on 15.2.2001*
245.	Ashok Ashram, Dehradun	Residential School (P)	5,54,030	13,40,190
246.	Purwanchal Jana Seva Sansthan, Ajamgarh	Mobile Dispensary	—	4,80,008
247.	Samajik Nyay Awam Manav Kalyan Samiti	Residential School	—	7,60,445
248.	Hindu Muslim Ekta Awam Kalayan Samiti, Lakhimpur Khiri	Residential School	—	7,60,845
249.	World Welfare Reportory, Lucknow	Residential School	—	8,46,346
250.	Janakalyan Awam Nari Uthan Samiti, Faizabad	Mobile Medical Unit	—	61,020
251.	Saket Mahila Mandal Kalyan Samiti, Gonda	Residential School	5,67,810 4,53,310	—
Meghalaya				
252.	R.K. Mission Chera Bazar	High School and Technology	8,58,319	
		46-Schools (P&M)	5,00,380	
		2-Hostels		1.06,39,526
		Dispensary		
		Library		62,00,786
		Mobile Dispensary		1,23,159
				10,73,435
253.	R.K. Mission, Shillong	Bus Hostel	4,35,990	4,62,220
		Library	4,35,990	5,11,400
254.	Sewa Bharati, Shillong	Residential School		7,44,210
				4,52,610
Sikkim				
255.	Muyal Langa Trust W. Sikkim	Residential School		7,59,960
256.	Human Development Foundation of Sikkim, Gangtok Sikkim	Residential School		7,73,227

Impact of NGO is Work on People: There is no doubt that there are a number of dedicated, committed and honest National level NGOs, who have contributed a lot in the development and welfare of the people. But it is also true that a number of NGOs have become professionals. They lack dedication, commitment and honesty. As a result, they have defamed the name NGO. They are money minded and want to earn money through NGOs not to provide welfare services to deprived community. They have been black listed, too. They have defeated the very purpose of NGO. It has sent wrong messages among the people regarding the selfless honest service to deprived humanity by NGO. Thus, the impact of NGO work on people can be discussed under two broad headings—(i) positive Impact and (ii) Negative impact.

Positive Impact of NGO's Work

The positive impact of NGO's work comes in to existence by the satisfactory work they have done from the funds received by the Government for welfare and development of SCs, STs, children, women, nutrition, health and sanitation. The positive impact of NGO's work on the people can be discussed as below:

1. NGOs have made valuable contribution towards the education of children of SCs, STs, minoriteis and other weaker sections of the society.
2. NGOs have provided training to SCs, STs, minorities etc in the field of agriculture and allied work.
3. NGOs have established training centres for providing training in animal husbandry, poultry, fishery, goatery, pigery etc.
4. NGOs have provided health services to the people by establishing 10-beded hospitals and mobile dispensary.
5. NGOs have provided people health education and value of safe drinking water for the health.
6. NGOs have made people aware of value of nutrition in the maintenance of health.
7. NGOs have made valuable contribution towards sanitation awareness among the people.
8. NGOs have made people aware towards the health of women and children.
9. NGOs have contributed towards empowerment of women, SCs STs and minoriteis

10. NGOs have provided training to SCs, STs and minorities children in computer, typing and shorthand.
11. NGOs have provided training in weaving, sewing, embroidery and crafts.
12. NGOs have contributed towards leagal redressal of cases of SCs, STs and minorities.

Negative Impact of NGO Work

1. Now-a-days, there is mushroom growth of NGO in our country. A good number of them do not have feeling of service to humanity.
2. A number of NGOs lack honesty, dedication and commitment towards the cause of deprived humanity.
3. A number of NGOs have been blacklisted by the Government because they had grabed the funds in their own pockets without doing welfare and development work.Thus, they were found guilty of financial irregularities.
4. Incidence of gifting or presenting NGOs to kins, relatives, friends etc. by bureaucrats, ministers and politicians has come in to existence. This has created dots on the neatness of NGOs and their work
5. Some big NGOs get the work done through small NGOs as contractors, who lack dedication and honesty. Their main aim is to earn money in the name of NGOs
7. The incidence of Professionalism has also appeared related to the working of NGOs.
8. There is also corruption in granting funds to NGOs by the ministry and bureaucracy. Even non-reputed NGOs are able to get more funds on the basis of bribery.
9. People have now started pointing fingers towards the NGO authorities who lives in urban areas, wear costly dresses, travel through own cars, organise meeting in costly hotels, and consume costly food items in the name of sceminar symposium, workshop, discussion and other.

9

International Agencies of Development

Development of people of a nation is not only responsibility of a National and State Governments and NGOs. But it is also a responsiblity of international Agencies and organisations. Although our country India is believer of International brotherhood (*Vasudhaiwa Kutumbakam*), but till now India has to fight five wars which have been imposed upon her by Pakistan and China time to time. At International level, too, two great world wars were fought. These two great world wars and other wars between countries caused great loss to International communities and the concept of International brotherhood had to receive a great jolt. But in spite of it, attempts have been made by scholars, philosophers and politicians to inculcate the feeling of International brotherhood among international communities of the world. In order to achieve this goal, emphasis was placed upon the formation, of International Organisations by the world communities. It was believed that such International organisation will be fruitful for peace, prosperity and development of world community in general and oppressed humanity in particular. But in spite of all such theoretical prepositions, International organisation could not be established before World War-I (1914-1918). In human and cruel experiences of World War-I, forced the famous politicians of those days to formulate such a organisation at International level so that world war could be stopped in future. After the end of World War-I, League of Nations was formed in 1919. But League of Nations became failure in achieving its goal and Second World War broken out in 1939. The cruel and inhuman impact of World War-II is still a reality for the people

Heroshima and Nagashaki. After World War-II, United Nations Organisation was established in 1945 for international peace and prosperity.

League of Nations

The impact of world war-I (1914-18) was really very terrible. Lakhs and lakhs of lives and millions and millions of properties were destroyed. Therefore, after the end of war, attempt for peaceful agreement was made. As a result of which league of Nation was established in 1919. In the beginning the membership of League of Nations was restricted only up to friend nations and those neutral nations. Thus, in the initial stage only 42 countries were the members of League of Nations. Later on, membership of nations rose up to 60. According to the rule of membership of League of Nations, an independent nation must have support of one-third members. On the basis of support of one-third members, any new independent country could be included in the list of membership.

Main objectives of the League of Nations: The following were the main objectives of the League of Nations:

1. Peaceful solution to international conflicts and wars.
2. Establishment of International peace and to check wars in future.
3. To enhance reciprocal cooperation in social, cultural and economic fields between countries of the world.

Functions of League of Nations: The main function of League of Nations was to safeguard Internal peace by ending reciprocal battle between two countries. But League of Nations became inefficient in this regard and could not achieve its success. However some success was achieved by the League of Nations. League of Nations was successful in tackling the war between Swiden and Phinland, in 1920 over the issue of Holland Island. Similarly, League of Nations achieved success in curbing wars of Sylesia, Greece, Bulgaria and Puruvia Lativia. But it could not succeed in solving the problem of Japan and

China over Manchuria. China and Japan both were members of League of Nations. But Japan made an attack on China in 1931. An appeal was made in League of Nations on this issue, but League of Nations could not do anything. The incidence of China proved that League of Nations was a weak institution. It could not do any harm to aggressive countries. It remained deaf spectator in other wars too. In 1935 Itally made an attack on Obecenia and captured it. Itally and Obecenia both were members of the League of Nations. But it could not prevent Itally to do so. It encouraged facist Mousolini. It also encouraged Nadi of Germany Hitler who made attack on Austria and Chekoslovakia and brought them in her territory. Hitler also attacked on Polland in 1939. This incidence led to the beginning of Second World War.

Welfare Activities of League of Nations: There is no doubt that League of Nations could not achieve success in political fields, but it could achieve its success in non-political fields like welfare activities. It performed a number of welfare works. It helped developing countries in amelioration of their economic conditions. It improved means of transportation. It extended help in scientific researches for human communities. It did praiseworthy work in controlling epidemic and calamities. In the eradication of slavery, sale and mortgage of women and exploitation of minorities, it played significant role. It could establish reciprocal cooperation in social, cultural and economic fields among countries of the world.

Causes of Failure of League of Nations: International communities were happy with the establishment of League of Nations. In fact International communities had great hopes from it. Its establishment had sent messages among the people regarding the commencement of Internationalism. But people were more loyal to their country than the International community. Loyality towards own nation at the cost of other nations made this institution very weak. The main causes of failure of league of nations could be listed as below:

1. As a result of First World War a number of problems arose which was difficult to solve by League of Nations.

2. Member countries lacked interest in objectives and ideals of League of Nations.
3. The member countries also lacked the feeling of sacrifice and cooperation.
4. Some powerful countries were not its members. For example, Soviet Russia was not its member from the beginning. America had also isolated itself from it in the beginning. Thus, it could not get cooperation of some powerful countries.
5. Heavy punishment in cash was imposed upon Germany.
6. League of Nations was just like a cooperative society.
7. After World War-I, dictatorship developed in Itally and Germany. It was not ready to accept the objective and ideals of League of Nations.
8. League of Nation could not establish peace in the world and Second World War broke out in 1939.

UNO (United Nations Organisation)

United Nations Organisation is in fact an association of 191 nations at present. The last nations to join this International Association are Switzerland and East Timor. These Nations being members of this association have pleaded themselves for the maintenance of international peace and security. The states as members of the association have agreed to cooperate in promoting social, economic, political and cultural ties to achieve the objective behind the formulation of this International Agency of development, welfare, peace and security.

Origin: The impact of Second World War was very terrible in comparison to World War-I. The people of Hiroshima and Nagashaki are still facing its terrible impact in the form of dum, deaf and disabled. To see the terrible impact of the war, Leaders of the world began to think and work in the direction of international organisation to control the wars in the world in future. The representatives of the USA, the United Kingdom, the USSR and Kuomintang China met at Dumbaraton in October 1944 to discuss the issue of formation of International organisation for peace, security, development and welfare of the world. The charter of United Nations Organisation was signed at a meeting of the representatives of 51 states held at San Francisco on June 24-25, 1945, representing two-third of the world population. The name of United Nations was given at the instance of President Roosevelt. United Nations Organisation actually came into being on 24th October 1945.

Objectives of the UNO: The objectives behind the formulation and formation of the UNO were as follows:

1. To maitain international peace and security.
2. To develop cooperation and friendly relation among nations of the world.
3. To promote and encourage respect for human rights and for fundamental freedom.
4. To solve international problems of human, social, economic and cultural character.
5. To employ international machinery for promotion of economic and social advancement of all people.

The purposes of UNO can be divided into four groups, viz., Security, Justice, Welfare and Human Rights. The members of this organisation shows commitment towards discharging their duties in these directions. Disarm, decolonise and develop are the three new objectives set forth by the United Nations.

The Headquarters of UNO: The headquarters of UNO was completed in 1952 at New York. The first meeting of General Assembly was held there in October 1952. The secretariate building stands on 17 acre tract of land donated by John D. Rockefeller on the Manhattan island, a suburb of New York. It has 39 stories and accommodates over 8,000 office workers.

Languages: The official languages of UN are English, French, Chinese, Russian, Arabic and Spanish. The working languages are English and French only.

Flag of UN: The Flag of UN is white on a light blue back ground with an emblem of two bent olive branches open at the top and in between them is the world map.

Membership of UN: Any sovereign nation who is ready

to accept the charter of UN can be its member. In the beginning, it had only 51 members, but now it has 191 members. India is one of original members of UN.

Organs of UNO: The UNO has following main organs:

1. General Assembly
2. Security Council
3. Economic and Social Council
4. Trusteeship
5. International Court of Justice
6. Secretariat.

1. General Assembly: It is the main body of UN. It may discuss any matter within the scope of charter except those already referred to the Security Council. The assembly discusses ways and means of preserving peace, disarmament, peaceful uses of atomic energy, economic development, social progress, non-self-governing colonies, human rights, etc. Its decisions are not legally binding.

It consists of all the member nations. Each nation can send five representatives, but each nation has only one vote. It meets regularly once a year. Special and emergency meeting can be summoned at the behest of the Security Council or otherwise.

It elects ten non-permanent members of the Security Council, all fifty four members of Economic and Social Council certain members of Trusteeship Council and Judges of International Court of Justice. The president of General Assembly is elected for a year.

The assembly initiates studies and makes recommendations to member states and to other organs for promoting international cooperation in political, economic, social, cultural, educational and health measures. It has power to discuss the working and functions of other organs.

It also passes the entire budget of United Nations. Thus, General Assembly is supreme supervisory organ of UN.

2. Security Council: It is the most powerful organ of UN. If general assembly is the supreme supervisory body of the world, Security Council is the most effective body of the world. It is peace maker as well as police man at the same time. It is peace maker because responsibility to maintain peace in the world is owned by it. It is policeman because it prohibits those who attempt to disturb peace in the world by using its means.

In the beginning Security Council had 10 members, in which there were five permanent members each from USA, Russia, France, England and Peoples Republic of China. Other 6 temporary members were elected by General Assembly for two years. On 1 January 1966, the number of members of UN Security Council was increased to 15. Out of 15 members, 5 are permanent and 10 are temporary members. All five permanent members are same and all ten temporary members are elected by General Assembly for two years. Out of ten temporary members, distribution was as follows:

Afro-Asian Nations	5 members
East Europe	1 member
Latin America	2 members
West European and other nations	2 members

A decision on peace and security requires the agreement of at least nine members, out of 15 members of Security Council. It can tell the friendly nations to solve their conflict in peaceful manner. But if they are unable to solve their problem, they can send the matter in UN Council. The decision of UN Security Council requires no negative vote from anyone of five permanent members. On getting positive vote of permanent members, it can recommend even for armed action for maintaining world peace and security. The aggressive nation can be punished in several manner. It does not have army of any kind. But it can take help of member nations for armed action. The permanent members have veto power. Member nations are not bound to obey the decision of General Assembly, but they are bound to obey the decision of Security Council.

The Security Council remains in session round the year. It can meet at 24 hours notice. The president of Council remains in office for one month only. He is succeeded in alphabetical rotation by the representative of the member countries of Security Council.

3. Economic and Social Council: Now-a-day, it consists of 54 members elected by the General Assembly. But in the beginning, it had only 18 members, which was rose to 27 members. Now it is 54 members council to provide greater representation to the various parts of the world. The members are elected for a term of three years. Through this council attempts are made to achieve international cooperation by solving intranational problems of social, economic, cultural and humanitarian character. It also promotes respect for human rights and fundamental freedom for all without discrimination of race, sex, language, religion, etc. It seeks to build a world of greater prosperity, stability and justice. It meets at least thrice in a year.

The main task of the council is to knit together the activities of specialised agencies. The charter provides for special commission in economic and social fields and also for promotion of human rights. The various commissions set up by the UN Economic and Social Council are as follows:

(i) Human Rights Commission
(ii) Economic Development and Employment Commission
(iii) Narcotic Drugs Control Commission
(iv) Woman Commission
(v) Population and Transport Commission

The UN Economic and Social Council works under the guidance and direction of General Assembly and supervises the work of specialised agencies.

4. ***Trusteeship Council***: This UN Council bears prime responsibility to supervise the administration of Trust territories which were placed under International Trustee system. It consists of member states administering trust territories and permanent members of Security Council not administering territory. The present administering state is USA (Trust Territory of Pacific island). The non-administering members are China, France, Britain and Russia. There were 11 original trusteeships. Most of them were in Africa. They attained self-Government either as separate states or joining neighbouring independent states.

5. ***International Court of Justice***: This court serves as principal Judicial organ of the UN. Its headquarters is situated

at Heg in Polland. It consists of 15 Judges elected by the Security Council and General Assembly. Five judges retire at every three years. Cases are decided by a majority of judges present. All member states which are parties to a dispute have to comply with its decisions. The court also decides the amount of compensation to be paid to those who are disabled or killed while performing their duties under the order of United Nations. All UN member nations are party to the state of the court. Earlier Dr. Nagendra Singh of India was elected a member of International Court of Justice. The vacancy caused by his death was filled by Mr. R.S. Pathak, former Chief Justice of India.

6. Secretariat: The UNO has its own secretariat. It is the chief administrative wing of the UNO. It has a working force of over 10,000 men and women of UN member nations employed there. The duties and responsibilities of Secretariat are exclusively international. It is headed by the Secretary General. Secretary General is appointed by General Assembly on the recommendation of Security Council. He is the chief executive officer of United Nations. Kofi Annan of Ghana was also Secretary General. His term had expired on December 2006. Ban Ki Moon is present Secretary General.

Successes of UNO

The chief successes of UNO are as follows:

1. Setting up of the Jewish state Israel (1948).
2. Establishment of the Republic of United States of America.
3. Ceasefire between India and Pakistan on Kashmir border (1949).
4. Driving away the aggressors in south Korea.
5. Specialised agencies have done good job in fighting poverty, ignorance, malnutrition and disease.
6. International convention on crime of genocide was approved by Assembly council. Genocide means deliberate and forceful destruction of a human group clan or tribe.
7. Universal declaration of Human Rights (1948),
8. Concrete plan for world cooperation to develop atomic

energy for peace. The International Atomic Energy Commission has been set up.

9. Withdrawal of Anglo-French and Israeli forces from Egypt, 1956.
10. Induction of UN Emergency Force in Egypt
11. UN specialised agency extended assistance to the member nations by sending experts to teach, to train and to help.
12. In the economic field, help to provide the ground-work upon which sound regional and national policies could be based.
13. The UN played significant role in Kango by the employment of civilians for its economic development and employing force to setup a united Central Government.
14. The peaceful transfer of west Iran to Indonesia took place on 1 May 1960.
15. Liquidation of colonialism from Asia and Africa.
16. Partial nuclear Test ban treaty among USA, UK and USSR was signed on 7th August 1963.
17. India and Pakistan cease-fire (1965).
18. Imposition of economic sanctions and diplomatic buy cott of South Africa.
19. Non-Proliferation treaty was signed by more than 100 nations.
20. Special session on Indo-Pak war of 1971.
21. Creation of United Nations Emergency Forces in 1973. The force consist of 7,000 men.
22. Conference of decertification, 1977.
23. United Nations Conference on Trade and Development.
24. Special sessions of the UN on energy crisis, population control, food problem, environmental pollutions etc.
25. Observing 1975 as Women's Liberation Year.
26. The year 1979 focuses on problem of children.

Failures of UNO

1. No agreement among big powers over atomic energy control.
2. No final solution to Indo-Pak dispute on Kashmir.
3. United Nations and world community could not curb

genocide.

4. Korea continued to be a divided nations.

Specialised Agencies of UNO

The UN has following specialised agencies and other bodies for discharging duties and responsibility:

1. ILO (International Labour Organisation)
2. FAO (Food and Agricultural Organisation)
3. UNESCO (United Nations Educational, Scientific and Cultural Organisation)
4. UNICEF (United Nations International Children's Emergency Fund)
5. WHO (World Health Organisation)
6. ICAO (International Civil Aviation Organisation)
7. IBRD (International Bank for Reconstruction and Development, or World Bank)
8. IMF (International Monetary Fund)
9. IFC (International Finance Corporation)
10. IDA (International Development Association)
11. IAEA (International Atomic Energy Association)
12. ITU (International Telecommunication Union)
13. UNCTAD (United Nations Conference on Trade and Development)
14. Conference on Human Environment
15. UNHCR (United Nations High Commission for Refugees)
16. UNIDO (United Nations Industrial Development Organisation)
17. ESCAP (Economic and Social Commission for Asia and Pacific)
18. UNFPA (United Nations Fund for Population Activities)
19. ADB (Asian Development Bank)
20. RCS (Red Cross Society)
21. CARE (Cooperative for Assistance and Relief Everywhere)

ILO (International Labour Organisation)

It was constituted in 1919 under the terms of Treaty of Versatile as an independent organisation of League of Nations. It became a specialised Agency of UN in 1946. It consists of three organisations which are as follows:

(i) General Conference. It consists of national delegations comprising two Government delegates, one delegate representing the employers, and one representing workers. The chief function or General function is to formulate international social standards in the form of conventions.

(ii) Governing Body. It is composed of 56 members, 28 representing Government, 14 representing employers and 14 representing workers.

(iii) International Labour Office: It provides the secretariat of the conference and of the governing body.

The headquarters of the ILO is situated at Geneva. ILO aims at improving labour conditions, raising living standards, and promoting economic and social stability. Beside, it collects and distributes informations on all subjects within the competence of the organisation.

It was awarded Nobel Peace Prize of 1969.

2. F.A.O (Food and Agricultural Organisation)

It was founded on October 16, 1945. The objectives behind the establishment of this organisations were as follows:

1. To help nations in raising their living standards.
2. To improve the condition of people in rural areas.
3. To improve quality of food.
4. To help people of the world in their continuing fight against hunger and malnutrition.

FAO has succeeded in improving the supply of food grains and rice. It has saved millions of lives from starvation. Efforts have been made to increase supplies of agricultural implements, fertilizers, seeds on a worldwide scale. Vaccination of cattle diseases have been made available. Alternative means of nutrition has been discovered by it. Since 1960, FAO has launched a worldwide Freedom from Hunger campaign. The World Food Conference was held at Rome in 1947 in which a number of long term decisions were taken in order to solve the food shortages.

FAO consists of the General Conference, the council and the office headed by Director General. The General Conference is composed of one representative from each of the member countries. It meets every year to determine the policy and approve the budget and world programme of FAO. The council has 50 member nations elected by the conference. It serves as the governing body of the FAO between sessions of the conference. Its headquarters is at Rome, Itally.

3. UNESCO (United Nations Educational, Scientific and Cultural Organisation)

Its headquarters is at Paris. It consists of two parts—(i) General Conference and (ii) Executive Board. The General Conference is composed of representatives of all member states. The Executive Board consists of 45 members elected by the General Conference. It executes the programme adopted by the Conference.

The main purpose of UNESCO is to promote peace and security in the world by cooperation among nations through education, science and culture so that people may have respect for justice, rule of law, human rights and no discrimination based on race, caste, creed, sex, language, region and religion. The USA had left UNESCO in December 1984, but has rejoined again in September, 2003.

The following are the main functions of UNESCO:

I. It helps in translating rare manuscripts.
II. It is a world centre of information for study, travel and work abroad.
III. It collects and disseminates scientific knowledge so as to reduce racial prejudices. It gives special

encouragement to scientific research.

IV. It promotes favourable conditions for intellectual cooperation among artists, musicians, philosophers and men of letters.

V. It suggests measures for preserving the national monuments, rare manuscripts etc.

VI. An International convention has been drawn up so that places of archaeological interests should not be damaged in the situation of international conflict.

VII. It carries a campaign against racial prejudices. It organised an International Round Table on Jawaharlal Nehru's role in modern world in New Delhi (September 1966). It was the first conference of its kind in the memory of great leader Pt. Nehru who promoted the idea of peace and friendship among the nations.

4. UNICEF (United Nations International Children's Emergency Fund)

The organisation was established by UN General Assembly on April 7, 1948. Its main purpose was to provide massive emergency relief to destitute young victims of Second World War. It is making available to millions of children every year milk, food and other nutritional materials. It assists in providing trained nurses, and midwives for maternity aid. It has helped a number of nations to launch sanitation projects. The most important of them is the provision of safe and clean drinking water for both rural and urban communities. In our country ICDS programme is being run with the assistance of UNICEF. Our country has received billions of rupees as assistance from UNICEF for equipment of health centres, health sub-centres, referral hospitals, 8 beded hospitals, public health laboratories, poultry and fishing farms. In 1977 and 1978, UNICEF provided assistance to the food and cyclone victims in India and Pakistan. Its source of income is voluntany contributions of Governments. In 1978, income for short and long term projects approached $120 billions. The Nobel peace prize for 1965 was awarded to UNICEF. In 1990, the UNICEF was awarded Rs.15 Lakh Indira Gandhi prize for peace and development. It was received at New Delhi by Mr. James Grant, Director General, UNICEF.

The words 'Intenational' and Emergency have been dropped from its name. Now it has been renamed as UNCF (United Nations Children's Fund).

5. WHO (World Health Organisation)

This International Agency of development was established on April 7, 1948. Its purpose was to assist Governments in raising standard of public health and to disseminate information on medical researches. It extends help to Government in breaking the vicious circle, i.e., sickness breeds poverty and poverty breeds sickness. Its some programmes are related to maternal and child health, nursing, blindness, tuberculosis, maleria, kalajar, cholera, AIDS, sanitation etc. Its headquarters is at Geneva (Switzerland).

The chief organs of World Health Organisation are as follows:

I. World Health Assembly
II. Executive Board
III. Secretariat.

WHO mobilised international campaigns against cholera, tuberculosis, maleria, AIDS etc. It has organised itself on the basis of regions. At present it has six regional organisations. In order to encourage medical researches, WHO has arranged for the investigation of cancer and heart diseases in different countries. It has organized an international network of laboratories which identify disease organisms, improve vaccines and train research workers.

WHO is now engaged in combating the menace of AIDS. According WHO estimate, AIDS affected people are nearly 40 millions. Its major responsibility is to make the world disease free. WHO extends full cooperation to the victims of natural calamity such as earth quakes, Cyclones, flood. It extends every support to check the break out of epidemic. April 7, is celebrated as the World Health Day every year. WHO has come into action to fight SARS (Severe Acute Respiratory Syndrome). This epidemic has been controlled. Thus, World Wide Calamity has been averted.

6. ICAO (International Civil Aviation Organisation)

It was established on 4th April, 1947. Its aim is to develop the principle and technique of International Air Navigation. Its objective is also to regulate International Air Transport. ICAO encourages the use of safety measures, uniform regulations for operation and simpler procedures at international borders. It promotes the new technology and methods. It has simplified Government customs, immigration and public health regulations. Its headquarters is at Montreal (Canada).

7. IBRD (International Bank for Reconstruction and Development, World Bank)

The International Bank for Reconstruction and Development, popularly known as World Bank, owes its birth to the deliberations of United Nations Monetary and Financial Conference which met at Brettonwood, New Hampshire, to prepare the final text of the Articles of agreement of International Monetary Fund and International Bank of Reconstruction and Development from July 1-4, 1944. The World Bank was established on December 25, 1944, when the articles of Agreement of Bank were ratified by requisite number of member governments. The second global war had completely dislocated the multilateral trade and had caused massive destruction of life and property. Economies of England and other countries in Europe had completely shattered. Therefore, need for promptly reconstructing the war damaged economics of England and other countries was recognised. It was also recognised that the stable world peace was threatened due to disparity in standard of living between developed and undeveloped countries. The need to raise the standard of living of people of undeveloped countries was emphasized in the Brettonwoods Conference. This ultimately led to the establishment of International Bank for Reconstruction and Development.

The main purpose of the World Bank is to promote economic and social progress in developing nations. It is an international corporate institution whose capital stock is owned by its member nations. The principal purposes of World Bank are as follows:

I. To assist in the reconstruction and development of territories of its member governments by facilitating investment of capital for productive purposes.
II. To promote foreign private investment by guarantees of or through participation in loans and other investment of capital for productive purposes.
III. Where private capital is not available on reasonable terms to make loan for productive purposes out of its own resources or out of funds borrowed by it.
IV. To promote the long range growth of international trade and the maintenance of equilibrium in balance of payment, of members by encouraging international investment for development of productive resources of members.

Headquarters of the World Bank is at Washington D.C. The World Bank's loan are directed to help members build foundation of sound economic growth. The Loans guaranteed by World Bank are for the purpose of specific projects of reconstruction or development. The World Bank ensures that loans sanctioned are used only for the purpose for which the loan is guaranted. The bank lends to only credit worthy borrowers. Loans are sanctioned for only those projects which are expected to yield high rate of economic return to the country. About 16.5 per cent of World Bank's total cumulative loans have been sanctioned for the development of electric power and other energy, 14.8 per cent for development of transportation, 16.5 per cent for agriculture, 6.2 per cent for industry, 10.6 per cent for finance, 4.9 per cent for education, 4.5 per cent for urban development, 4.4 per cent for water supply and sanitation, 9.2 per cent for multi-sector. 3.7 per cent for oil and gas, 2.0 per cent for population, health and nutrition, 2.0 per cent for public sector management and the balance 4.7 per cent for social sector, telecommunications, environment, mining and other extra active and urban development projects.

Any country is eligible for the membership of the World Bank, if it subscribes to its charter under the Bank's Articles of Agreement. Only those countries who are members of International Monetary Fund (IMF) can be considered for the membership of the World Bank. Subscription by member countries to the capital stock of the World Bank are related to each member's quota in IMF, which reflects the country's relative economic strength. The total number of members of World Bank is 180 as on June 30, 1997. A member can withdraw anytime its membership from the World Bank. Its withdrawal is effective from a written notice issued to the member by the World Bank. Failure to fulfil its obligations towards the Bank may lead to suspension of a member. Even when a Government ceases to be Bank's member, it is obliged to repay on demand its portion of losses.

The Bank has four significant parts, which are as follows:

I. Board of Governors
II. Executive Directors
III. President
IV. Staff

All powers of Bank are vested in the Board of Governors consisting of one Governor and one alternate Governor appointed by each member for five years. Alternate Governor cannot vote, except in absence of the principal. Each Governor has the voting power which is related to the financial contribution of the Government which it represents. The USA has 17.52 per cent of the total votes and the UK has 4.9 per cent of the total votes. The Board of Governors meets once every year. Informal exchange of views on major international financial and monetary problems takes place in the meeting of the Board of Governors besides dealing with matters of formal action.

There are 24 Executive Directors to direct general operations of the World Bank. Five Executive Directors are appointed by five biggest shares—USA, UK, Japan, Germany and France. The remaining 19 are elected by the remaining members. Each director holds voting power in proportion to the shares held by his Government. With certain exceptions the Board of Governors has delegated all its powers to the Executive Directors who are responsible for the conduct of the general operations of the Bank. The Executive Directors function in continuous session and meet regularly once a month. A majority of the Executive Directors exercising 50 per cent or more of total voting power constitutes a quorum. The President of the Bank acts as the chairman of the Board of Directors. He has no vote except a deciding vote in case of an equal division. He is the chief of the operating staff of the Bank and is responsible to the Board of Governors of the Bank for conduct of ordinary business of Bank and its organisation. He is assisted by a number of departments with each department under the charge of a separate departmental head.

The Bank sanctions loans to members in anyone or more of the following ways:

I. By participating in direct loans out of its own funds.
II. By granting loans out of funds raised in the market of a member, otherwise borrowed by the bank.
III. By guaranteeing in whole or part loans made by private investors through the investment channels.

The Bank also provides assistance, in establishing strong national institutions to promote development. It also provides technical assistance in development programmes through various survey missions. The survey mission makes intensive studies of national resources of developing member countries and makes recommendations for long term development programmes.

Regionwise distribution of Loan and Amount involved in the fiscal year ended on 30.6.1997.

Region	*No. of Loans*	*Amount (US$ Millons)*
1. Africa	02	50.0
2. East Asia and Pacific	27	4,074.4
3. South Asia	04	626.5
4. Europe and Central Asia	47	4,560.9
5. Latin America and Caribbean	48	4,437.5
6. Middle East and North Africa	13	769.6
Total	141	14,524.9

Ever since its inception, the World Bank has lent a cumulative total amount of $295,267.7 million up to June 1997 in the form of 4,064 loans mostly to the developing countries.

The Bank supported technical assistance in early years and in 1970s and 1980s had focused on engineering assistance in designing bridges, dams, highways, telecommunication systems etc. In recent years, technical assistance has been directed to capacity building, a complex process of creating and disseminating knowledge for development purposes at all levels of society. It is largely culture bound and process oriented. For this purpose Institutional Development Fund was (IDF) was established in 1993. It has made 335 Grants to 108 member countries. The Bank has close working relation with United Nations Development Programmes (UNDP). During 1996-97, Economic Development Institute (EDI) held more than thirty governance related activities in more than a dozen countries across the five regions.

In 1996-97 the Bank with IMF look initiative in providing relief to the Heavily indebted poor countries (HIPCs) by reducing their external debt burden by initiating HIPC (Debt Initiative Trust Fund).

On March 29, 1965, IBRD had established a machinery to settle disputes between member nations and foreign investors. The machinery known as convention on settlement of Investment Disputes between states and nations started its operation on October 14, 1966. Under the convention, an International Centre for Settlement of Investment Disputes (ICSID) was established in 1966. By June 1997, 127 member countries had ratified the convention. The ICSID publishes an annual law journal.

The Bank also plays role in promoting international peace by resolving successfully difficult international dispute. Bank activities have been diversified by including urban development, population planning and tourism within its activities. A population projects Development and a population studies Division have been set up in Economic Department of the Bank. It has also established a new Tourism Project Department. It provides technical assistance to the International Finance Corporation on possible new IFC tourist investment. Banks tourism mission have visited many countries.

In addition to the conventional loans, World Bank has made sincere efforts to secure outside assistance from developed countries for underdeveloped countries. Due to the sincere efforts of the Bank, a consortium of 12 lending western nations known as Aid India Club comprising UK, USA, Germany, Japan, France, Canada, Italy, Sweden, Austria, Belgium, Neatherlands and Holland was formed to help India out of her foreign exchange difficulties. The Aid India Club had provided $ 5,472 million financial assistance to India during Third Plan Period to extend help in India's economic development programme.

The India Development Forum (IDF) is a consortium of international donors had pleaded further a total assistance of $ 66.9 billion to India during the fiscal year 1988-89 to 1997-98. The Bank had also established help Pakistan club.

Thus, the World Bank has played significant role in development of the world in general and developing countries in particular.

In spite of its vivid plans for the economic and social progress of world in general and developing and under developing countries in particular, the World Bank has also been subjected to criticism. The main criticism labelled against the World Bank are as follows:

1. The World Bank charges a very high rate of interest.
2. The World Bank applies orthodox standard of judging the transfer capacity of the borrower country.
3. The Bank does not give loans purely on merit and economic considerations in practice.

8. IMF (International Monetary Fund)

The United Nations Monetary and Financial conference of 44 nations was held at Brettonwoods, New Hampshire in USA in July 1-22, 1944. This conference gave birth to International Monetary Fund (IMF) and International Bank for Reconstruction and Development (IBRD) popularly known as World Bank. IMF started functioning from December 1945.

The purposes behind the establishment of this International Agency of Development were as follows:

I. To promote international monetary cooperation through a permanent institution.

II. To Facilitate the expansion and balanced growth of International trade.

III. To promote exchange stability, to maintain orderly exchange arrangements among members and to avoid competitive exchange depreciation.

IV. To assist in the establishment of a multilateral system of payments in respect of current transaction between members and in the elimination of foreign exchange restrictions which hamper the growth of world trade.

V. To provide confidence to members by general resources of fund made available to them under adequate safe guards.

VI. To shorten the duration and lessen the degree of disequilibrium in the international balance of payment of member.

The structure of IMF consists of following parts:

I. Board of Governors
II. Board of Executive Directors
III. Managing Director
IV. Staff

The highest authority of IMF is the Board of Governors, in which each member country is represented by a Governor and an alternate Governor. IMF Governors are generally Finance Ministers of the concerned countries or heads of Central Banks of the countries concerned. IMF Governors meet once in a year. They may meet or vote by mail at other time. There are 24 IMF Governors.

The Board of Governors has delegated many of its powers to Board of Executive Directors. The Board of Directors is responsible for conducting the business of the fund. The headquarters of the IMF is at Washington DC. Board of Governors remains in session at the IMF Headquarters. There are 24 Executive Directors. Of these 24 Executive Directors, 6 Executive Directors are appointed. 18 other Executive Directors are elected for two year terms representing the 175 members of the IMF. Out of six Executive Directors, five Executive Directors are appointed by members with largest quota—USA 17.78 per cent, UK 9.48 per cent, Germany 5.54 per cent, France 4.98 per cent and Japan 5.54 per cent. One Executive Director is appointed by Saudi Arabia 3.45 per cent by virtue of its being one of the two largest credit members.

The Executive Board elects the Managing Director, who is chairman of the Executive Board. In addition, he is the chief of operating staff of the IMF. Under the direction of Executive Board he conducts the ordinary business of IMF.

Besides sanctioning finances, IMF provides technical Assistance to its member countries. It is an integral part of IMF. IMF provides technical assistance in four broad areas which are as follows:

I. Designing and implementing fiscal and monetary policies.
II. Institution building like central bombs, treasuries, tax

and custom administration.

III. Collecting and refining statistical data.

IV. Drafting and reviewing financial legislation.

The technical assistance is provided by IMF through Monetary and Exchange Affair Department, the Fiscal Affair Department, the Statistics Department, IMF Institute, Legal Department and Treasure Department.

India is founder member of IMF. India has played important role in formulation of IMF Policies. India is grateful to IMF for coming to rescue in her period of crisis.

9. IFC (International Finance Corporation)

It was setup in July 1956 to encourage the investment of private capital in productive capacities in the less developed areas of the world. The chairman of the World Bank is the ex-officio chairman of the Corporation. Membership of IFC is open to those countries which are members of the World Bank. Headquarters is at Washington D.C.

10. IDA (International Development Association)

It was set up in 1960 to function as an affiliate to the World Bank at the initiative of the U.S Government and World Bank. It provides external development capital to countries which due to balance of payment reasons are unable to raise sufficient finances abroad on conventional term. Its headquarters is at Washington D.C. World Bank, IFC and IDA work in close collaboration and form a closely knit team.

11. IAEA (International Atomic Energy Agency)

It was setup under UN to make recommendations for the use of atomic energy for peaceful purposes. All nations are expected to place a part of fissionable material and also uranium at the disposal of the Agency. The main objectives of IAEA are as follows:

I. To accelerate and enlarge the contribution of atomic energy to peace, health and prosperity thrughout the world.
II. To ensure that assistance provided by it or at its request or under its supervision or control is not used in such a way as to further any military purpose. This Agency was established at a meeting of 80 nations held in New York in September 1956. Its headquarters is at Viena (Austria).

12. ITU (International Telecommunication Union)

This union was founded at Paris in 1865 as International telegraph Union. Its name was changed as International Telecommunication Union in 1934. The ITU is responsible for regulating, coordinating and planning all types of international telecommunications, including space radio communication. It allocates radio frequencies and maintains a register for all such frequencies for all purposes through out the world. ITU is particularly concerned with the development of new techniques such as space communication.

13. UNCTAD (United Nations Conference on Trade and Development)

A conference of about 100 nations was held in 1964 at Geneva to consider measures to improve the economy of developing nations. There is great disparity in standards between the developing and the developed nations. The former are anxious to improve living standard to speed up economic and social progress. Their needs for imports of development goods and the technical knowhow and foreign capital are increasing, but terms of world trade have operated to their disadvantage. The export prices of raw materials are declining and the prices of imported machinery and other products are increasing. To solve this problem of imbalance, the following suggestions were resolved:

I. The industrialised nations, viz., USA, Japan, Western Europe, Canada, Australia, should provide one per cent of their national income to the developing countries.
II. An International Trade Organisation consisting of 77 Nations comprising the development countries of Asia, Africa, Latin America to keep world trade problem under constant review and to promote an extension of international trade.
III. The developing countries demanded for preference in industrialised nations and 50 per cent cut on duties on certain types of goods was turned down.
IV. The landlocked countries demanded the freedom of rights of transit of their goods to the sea.

14. Conference on Human Environment

In 1968 UN General Assembly decided to convene a conference on Human Environment to seek solutions to many problems of pollution other than menances to health. Following extensive advance work, a conference was held at stock home in June 1972. The conference was attended by 1200 delegates from 10 nations. The conference prepared a blueprint for action at international level. The conference approved an action plan containing 106 recommendations. For measures to be taken up by the governments and international organisations to protect life, control contamination from man made pollution and to improve cities and other human settlements. The plan includes a global monitoring programme called **'Earth Watch'.** This programme was formulated to identify and measure environmental problems of world importance and to warn against impeding crisis.

The United Nations Conference on Human Settlement (HABITAT) was held at Vancouver from May 31—4 June, 1976. A plan of action to improve the quality of life in human settlement was formulated. This became an urgent need for accelerating urbanisation throughout the world.

United Nations Water Conference and united Nations Conference on desertification both were held in 1977.

World Environment Day is observed annually on 5th June.

United Nations Environment Programme (UNEP) has its headquarters at Nairobi. It is the first global United Nations Agency working in developing country. It stresses upon the conservation of forest, soil, water, air and sustainable development. The worst danger to the environment took place in Indonesia, Malaysia, and other South East Asian countries. The Major airports of the regions were closed due to forest fires in 1998. Ozone hole over the Antarctica and the emergence of Greenhouse effect is also a matter of concern for UNEP.

15. UNHCR (United Nations High Commissioner for Refugees)

It was established on January 1, 1951, by the UN General Assembly. It provides legal protection and material assistance to refugees. There are about two millions refugees of concern to UNHCR in Africa, Kampuchea, Pakistan, Thailand etc.

In May 1971, UNHCR's focal point was to assist ten million refugees of East Pakistan, now known as Bangladesh UNHCR had extended full cooperation to Indian Government in providing protection to ten lakh refugees of Bangladesh in 1972. It had owned Nobel Prize in 1954. Jeneva is its headquarters.

16. UNIDO (United Nations Industrial Development Organisation)

UNIDO came into existence in January 1967. Its purpose was to promote and accelerate the industrialisation of developing countries. Its headquarters is at Viena. The main functions of UNIDO are as follows:

I. To provide assistance to developing countries to expand, modernise and operate their industries including agro-industries.

II. To assist developing countries in obtaining external

finance for industrial projects.

III. To undertake operational activities including measures for effective application of modern method of industrial production.

IV. To coordinate all activities of United Nations related to industrial development.

The first conference of UNIDO was held at Viena in 1971. It was attended by 71 countries of the world. The second world conference was held at Lima, Peru in March 1975 in which it was resolved that the share of world industrial production should reach 25 per cent by the year 2000 AD against the then figure of 7 per cent. Nearby three quarters of global population lives in developing countries.

17. ESCAP (Economic and Social Commission for Asia and Pacific)

It was established by United Nations in March 28, 1947 under the name united Nations Economic and Social Council. It received its present name in 1974. This commission emphasizes upon social justice and integrated development.

18. UNFPA (United Nations Fund for Population Activities)

This organisation was created in 1967 to promote population programmes and extend systematic and sustained population assistance to developing countries. It helps them in dealing with their population problems. The UNFPA issued a very timely warning to India in population increase. Its headquarters is at New York.

19. ADB (Asian Development Bank)

The Asian Development Bank is a multinational development

financial institution. It was established for the purpose of lending funds, promoting investment and providing technical assistance. Its purpose is to promote the economic and social progress of its developing member countries: in Asia and pacific regions. The charter of Asian Development Bank was ratified by the majority of the signatory countries in early 1966. It started its operation in Manila in December 1966. The Bank operates under the agis of ECAFE (United Nations Economic Commission for Asia and Far East). Its headquarters is at Manila, Philippines. It is an Asian Bank conceived by ECAFE. Its capital is subscribed by its 40 regional members and 16 countries from outside the region. The membership of banks extends beyond region. Some countries outside Asia have contributed to the capital of the Bank. They are represented on the board of directors and professional staff of the Bank. There are 16 such non-regional members. It has a total staff of 1939 coming from 44 of the 56 member countries. Of this total staff 1939, 653 were professional staff and 1286 supporting staff.

20. International Red Cross Society

It is an International Agency of Development. It was promoted by war hours. The Geneva convention of 1964 laid down the principles assuming the safety of ambulances, hospitals, stores, and personnels distinguished by the emblem of Red Cross. The symbol was adopted by reviewing the order colour of Swiss Flag, which has a red back ground with white cross Besides dealing with associated problems of war, such as refugees and care of disabled persons, it also works for earthquakes, cyclones, flood and epidemic victims. It works in close touch with St. John Ambulance Association. Its main functions are as follows:

I. Promotion of health
II. Mitigation of suffering
III. Elimination of disease.

It has its organisations in different countries of the word.

21. CARE (Cooperative for Assistance and Relief Everywhere)

It is an international Agency of development. It was established in 1946. Its full name is Cooperative for Assistance and Relief Everywhere. Its main purpose is to promote mother and child health. In our country, CARE India has launched mother and child development programme. ICDS (Integrated Child Development Programme, NHNP (National Health Nutrition Programme) and Kishori Yojna are being run by CARE in our country. It also extends services during natural calamities like flood, earthquake, cyclone, etc.

10

Media as Agency of Development

Development is meant for the people but development news go to the people through media. Thus, media acts as agency of development. Media are regarded as fourth pillar of democracy. They act as carrier of news from the Government to the people and opinion, reactions and activities of the people to the Government. Thus, media are means of communication between the Government and the people. In our country majority of people live in villages. A good number of them are illiterate. The number of literates are more in towns, cities and metropolis. The towns, cities and mertropolis have better facilities of road, transport, electricity and communication. But villages are far away from s uch communicational facilities. Therefore, in our country one type of media cannot be a successful agent of development. Various types of media are required for the purpose of spread of development news in different parts of our country. Various types of media which play significant role in development of the people are as follows:

I. Folk Media
II. Mass Media
III. Electronic Media
IV. Print Media

I. Folk Media

In our country a number of communities reside in hilly, forested, desert, coastal and island areas. They are generally termed as Scheduled Tribes in our country. Again majority of people live in villages which are located in plain areas. The members of folk and peasant societies are illiterates in majority. There are also lack of roads, railways, and electricity. So, the people of these areas are generally dependent upon folk media. The folk media have following channels:

***1. Family Head*:** The head of the family can be used as a conveyer of development messages orally. Family head is authority of the family. Development related messages conveyed by the head has important bearing upon members of the family. Message related to education, nutrition health, sanitation, family welfare social, cultural, economic and other kinds of development become the speech event in the family. The member of the family have conversion on development issues among themselves. Only then they become ready to participate or not to participate in development programmes. Thus, the head of the family acts as one channel of folk media. Therefore, for effective implementation of development programme, heads of the family must be included in meeting, discussion and other things related to development news.

***2. Village Head*:** In tribal society of our country, there is a post of village head in each village. He is termed by different local terms, but he is most respected man in the village. His suggestion and decision are regarded by all. There is council of elders to extend help to the head. His words become speech events in the entire village. People gathered around the fire during winter season, in youth dormitory, in field where people go to work etc. If village head is included as one of channels of folk media for conveying messages related to development, the community people grasp those news positively and they become ready to accept and participate.

***3. Religious Head*:** In tribal villages, one finds the existence of a religious head. He termed as pahan or Baiga priest. *Pahan* or *Baiga* is respected by the community members. He transmits messages orally to the people. The people have full faith on news provided by the *Pahan*. A number of general and ritualistic news is transmitted through the *Pahan*. The *Pahan* can be used as messengers of development programmes. Tribal people receive the news communicated by the *Pahan* as cent-per-cent true because he is regarded as the representative of God and Goddesses in the community. The help of *Pahan* can be beneficial for the spread of developmental news in the community.

4. *Inter Village Head*: The tribals have inter village panchayat which has a head. This system is prevalent among tribals of Chotanagpur. It is known as Munda-Manki system, Paraha System, Pir System, Parganait System and others. Manki, Paraha Raja, Pir, Parganait etc. are respectable persons for the villages coming under a paraha. What ever they say, have impact upon the people. Manki acts as one of channels of folk media in transmitting various types of news among the people. If he is involved in communicating news related to family welfare programmes, literacy mission, education of girl child, maternity and child care, nutrition, health and sanitation and economic development programmes, the programmes will certainly receive good response of the people.

5. *Medicine Man*: Tribal villages, have a medicine man who diagnoses the diseases and provides herbal medicine to cure them. He commands respects in the tribal society. Words coming out of his mouth is taken as true. If he conveys news related to family welfare programme, nutrition, health, and sanitation programme and others, these programmes have better response by the people. This channel of folk media must be utilised for conveying developmental news among the people.

6. *Village Leaders*: In villages whether tribals or peasants, there are some village level leaders. They have contact with Government officials and political parties. Their words have command and respect in the society. The village leaders act as one of channels of folk-media. Their services can be utilised in the spread of development news among members of the village community. They can play significant role in transmitting orally the news related to development of village and people. The news through the channel of village leaders can bring more consciousness and participation of the villagers in development programmes.

7. *Weekly Haat*: Weekly haat is a significant feature of the tribal area. As it takes place once or twice in a week, that is why it is called weekly Haat. The day and time of weekly Haat remain fixed. In weekly Haat, males, females and children of neighbouring villages assemble for sale and purchase of the commodities. Here they also meet with Inter-village heads and functionaries. They get occasion for exchange of news with relatives, friends, inter village functionaries, shopkeepers and others. Thus, weekly Haat is very important channel of folk media. If a *vikas mela* (Development fair) is organised at weekly Haat and development news are conveyed to the people in local languages, it can bring more fruitful result. This channel of folk media must be employed as an agency of development in tribal areas.

8. Jatra Mela: In tribal areas, one finds the existence of Jatra Mela. This mela takes place at different places. In Ranchi, famous Jatra mela takes place in Murma, a place of origional settlement of Madira Munda after whom Nagvanshi raj was founded in the Chotanagpur. Jatra mela takes place after harvesting the paddy crops. This mela has an influence area. People from villages situated nearly 15-20 kms pay a visit at this mela site. This mela is visited by males, females, children, aged, youth and village and inter-village functionaries. Varieties of shops of sweets, commodities, agricultural implements, utensils, clothes, etc. are organised here besides different kinds of plays. Jatra mela acts as very significant channel of folk media. If vikas mela is also organised on this occasion by different development agencies, it can bring more development awareness and participation of people in development programmes of various kind.

9. Local Pilgrimage Centre: In tribal areas, there are many local pilgrimage centres where they go to worship. For example, in Jharkhand there are a number of such pilgrimage centres. For example, Rajrappa, Nagar Uttari, Tanginath, Chandawa, Itkhori, Bundu, Sun temple, Amareshwar Dham khuti etc. Local and regional people pay visit on these places on different occasions. A fair of devotees, visitors and sight viewers take place. These pilgrimage centres also act as channels of folk media. Organisation of vikas mela on the occasion of pilgrimage can bring more fruitful result of development programme.

***10. Folk Singers*:** Folk singers are important channels of folk media. They create songs on cultural heroes, heroines, social problems, national events, etc. They sing their song as well. They also move village to village and earn livelihood by singing songs, which they create. They receive in cash and kind as payment for singing their songs. They have manar, *Damaru*, *Kartal* etc. to play at the time of singing. They sing individually or in group. They are also found singing song in weekly Haat, Jatra mela and place of pilgrimage. The folk singers must be employed to create development related songs and to sing those songs at weekly Haat, Jatra mela, place of pilgrimage, villages etc. This will convey development news in song and sung form. It can make people more aware of development programmes.

***11. Folk Dramatists*:** In tribal villages as well as in peasant in rural areas, one finds the existence of folk-dramatists. They create drama on cultural heroes, heroines, social problems, national events, regional events etc. They stage their drama in a prose or poetry form. In local term they are called as Leela, Natak, Nautanki etc. They stage their drama by visiting village to village or on the occasions of great worship or festivals. Thus, folk dramatists act as channel of folk media. If these folk dramatists are asked to create development programme related drama in local dialict and to stage them in villages, they can be very effective in creating consciousness towards development programmes.

***12. Folk Artists*:** In folk and rural areas one finds the existence of folk-artists. They express their feelings in picture forms. They draw pictures and use them as code of non-verbal communication. Folk artists are also significant channel of folk-media. Folk artists can be utilised in drawing development related pictures and diagrams and to show them to folk and rural people. They can be helpful in creating development related consciousness. They can also be employed for wall drawing.

***13. Slogan Shouting and Procession*.** Slogan shouting and procession have bearing on the mind of the people. They create consciousness among the people. Thus, they act as important channel of folk media. Village level leaders, school teachers, Janasewak, compounder, nurses etc. can play significant role in making slogan in local dialect. They can organise processions with the help of village children. Slogan shoulting related to development programmes and organisation of procession can bring fruitful result of development programme. As folk idioms, and proverbs are used as language codes, they convey effective messages to the people.

14. Gram Sabha: Gram sabha can play significant role in formulation, implementation and monitoring of development programmes. It is playing significant role in rural development programme launched by the panchayat. Similarly, Gram Sabha can also be conveyed for other kinds to development programmes and selection of beneficiaries: The gram sabha should also include the Government officials like Janasewak, panchayat Sevak, BDO and Block doctors as representative of the Government. Gram sabha can be very effective medium of development programme.

15. Parents-Teachers Meeting: Now-a-days, almost every village has a lower school or primary school. These schools have Government paid teachers to teach children of villages. Parents and teachers both are integral parts of the folk media. Both should not act in isolation. They should come together for the education of village children. A regular parents teachers meeting will bring fruitful result in creating awareness towards development of education. They can also solve problems of children in school or in home. School teachers are important channels of folk media and they play significant role in educational development.

16. Janasevak: Janasevak is a Government official. He conveys developmental news to the villagers. He also makes villagers aware of different types of development programme going on through the Block and Gram panchayats. He also plays significant role in the selection of the development beneficiaries from among the villagers. Janasewak is a very important folk channel of development.

17. Swasthya Sevak and Sevika: In villages, there are health sub-centres. On these Health sub-centres, Swasth sewak and sevika are employed by the Government. Block doctors also pay a visit on these centres. They also organise swasthaya mela at the centres. Health sevak and sevika make a door to door visit for informations of diseases, family welfare, maternity and child care, nutrition, health and sanitation programme. Thus, swasthya sevak, sevika and doctor play significant role in conveying varieties of messages related to health development programmes.

All above mentioned channels of development are useful for conveying the developmental news among the people. If they can be utilised properly, they can definitely bring fruitful result.

II. Mass Media

Mass media also play significant role in the development of people. Mass media create awareness and consciousness among the people towards development programmes which have been launched by the Government. They also make people aware of processes of selecting beneficiaries and benefits to be achieved by participation. Mass media have their own channels of development. Important channels of mass media are as follow:

1. Radio
2. Television
3. Cinema

1. Radio: Radio is very important means of mass media. Radio listeners are found in almost all parts of our country: one finds the existence of radios even in those remote villages where one has to face great trouble in going and coming. There is no need of electricity for operating a radio. They are operated with the help of batteries. Radios are now available in various shapes and costs. They are easily available and can be kept even in pockets. That is why this have reached even in remote area. Our Government has Akashvani Kendras which are situated in different parts of the country. Several programmes are relayed from Akashvani Kendras. They also relay programmes in local or regional languages. People not only hear songs and national or state news, but they also hear various rural programmes which are relayed by the Akashvani Kendra. Radio channels also provide news of various development programmes which have been launched by the Government. Akashvani Kendra invite experts of rural, urban, health, education, nutrition, sanitation development. With the help of such experts they communicate developmental messages to the people. Girl education, boy education, maternity health, child care, general health, family welfare sanitation, nutrition, bonded labour, child labour, dowry, foeticide etc. related songs, slogans, conversation, drama and chaupal programmes are relayed by the radio channels. These programmes have impact upon the listeners residing in folk or urban areas. Developmental news becomes the subject of conversations in family and community. Through discussions and conversations people become conscious towards developmental programmes and issues.

2. ***Television***: Whereas Radio is audio channel of mass media, Television is audio-visual channel of mass media. The viewers of television channels and serials are more located in urban areas where there is continuous electric supply. The villagers who have electric supply also have televisions. Rural elites have Televisions. They operate Televisions with battery. Villagers gather at their doors to see TV Serials. TV attracted the rural masses of our country during the relay of *Ramayana* and *Mahabharata* serials by Doordarshan Kendras, which is a Government Channel. Doordarshan channels relay a number of serials and programmes related to development. People hear and see the speakers, actors, actresses, leaders, media persons, etc. on TV. Attitudinal change in masses have taken place towards the education of girls to hear and see programmes on TVs. Similarly, they have been attracted towards family welfare programmes, maternity, childcare, nutrition, health, dowry death, untouchability, child labour etc. after seeing and hearing programmes through TV Channels. Thus, the role of TV cannot be ignored in diffusing the developmental news and creating awareness towards development in masses.

3. ***Cinema***: Cinema is also one of the channels of mass media. Film producers have produced a series of films on social

evils and cultural taboos which have been detrimental for the development of the people. The evil practices of dowry, dowry death, bride burning, taboo on girls education, untouchability, exploitation of labourers, physical exploition of women foeticide evils of riots, benefits of educating girls, benefits of family welfare programmes etc. have been filmed and shown to viewers in cinema hall. Now-a-days, film cassets are available. They are seen on TVs with the VCD and VCR. Films produced on development policies, development programmes and their impact can be fruitful for the development of the people.

III. Electronic Media

Electronic media have been useful in sending and receiving messages from not only different parts of the country, but also from different parts of the world. Electronic media are also playing significant role in development of the people. The following types of electronic media are used for give and take of messages:

1. Landline Telephones
2. Mobile Telephones
3. Computer
4. Internet

1. Landline Telephones: Landline telephone services have been provided by Bharat Sanchar Nigam Limited as well as some private agencies. Through landline telephone, one can get message from any part of country as well as the world. Telephone services have lessened the distance and the time. One can receive messages from relatives, friends, schools, colleges, universities, banks, Government offices, hospitals, businessman, traders, tourist spots, hotels etc. Messages increase our knowledge and make us powerful. Thus, landline telephones are useful in communication development as well as general development of the people.

2. Mobile Telephones: Through mobile telephones one can keep the entire globe in his/her hand round the clock. He/she is connected to the people of entire world twenty tour hours. Useful messages on mobile phones can be obtained any time by the consumer because consumer keeps mobile phones always in his/her pocket. This telephone service has lessened the distance and time between friends, relatives, parent children, teacher-taught, doctor-patient, government-public, traders-consumers, wholesaler-retailer, implementing authority and beneficiaries. One can get messages from anyone or any institution located in any part of the world. Mobile phones have played significant role in the communicational development.

3. ***Computer***: Computer is also a significant means of development. It has brought knowledge revolution through information technology. Computer has been able to develop the concept of global village. Through computer a village has become an integral part of the world. Today one can find the existence of computer in railway station, bank, hotels, shops, trading centres, hospitals, telephone exchange, Government offices, school, post-office, college, Universities, and many other day-to-day activities. Though computer is operated by the people, but people are relying more on computers. One can get information from any other parts of the country and globe in few seconds. Computers are also helpful in typing, designing and other kinds of business and trade activities. Multinational companies are flourishing in different parts of the globe through computer and information technology. Thus, computer has become an important electronic medium of development.

4. ***Internet***: With the help of telephones and computers one is now able to receive messages through internet. Informations related to admissions in top level schools, colleges, Universities, engineering institutes, medical institutes, management institutes etc. are now available on internet. Informations related to railway reservation is also available at internet. Exam results are available on internet. Marriage mates are also available on internet service. Informations about development programmes can also be obtained through internet. Now every ministry of State and Central Government has internet service from where development related news can be obtained. Thus, internet as one of electronic media has played

significant role in development of people.

IV. Print Media

In our country print media are working for the development of the people and nation from the time of British period. Through their print news, print media have played significant role in achieving independence of the country. They are still playing valuable role in national reconstruction. Print media include the following:

1. *News Papers*: In different parts of our country various kinds of daily newspapers are published in Hindi, English and regional languages. These newspapers contains a number of news related to development programmes. They also present news about evaluation report of several programmes. Corruption, success and failure of the programme are also news items of daily newspapers. Those who are literate, they become acquainted with development news after reading the news papers. In order to know and understand the news, readers must possess knowledge of the written language in which newspapers are being published. Some important newspapers of national importance are Hindu, Times of India, Economic Times, National Herald, Hindustan Times, etc. In Hindi speaking belt newspapers are published in Hindi. For example, Prabhat Khabar, Dainik Jagaran, Hindustan, etc. Basides several news papers are published in Urdu, Bengali, Orria, Marathi, Kannad and other regional languages.

2. *Employment News*: A special newspaper dealing with vacancies in Government sector all over the country is published in a newspaper entitled *Employment News*. This newspaper is published weekly. This is very useful for youth and their career.

3. *Journals*: A number of journals are published by the editors of different institutions to highlight their news at national and international level. For example, Law Journal, Medicine Journal, Science Journal, Social Science Journal, Social Research Journal, Human Rights Journal, Economic and Political Weekly Health Journal, Women Journal etc. These Journals contain valuable articles written by scholars who have expertise in different development fields. A reader can acquire knowledge in different development fields after going through these journals. Journals are also published by the government offices and ministries to provide development news to the readers. Journals are published monthly, quarterly, half yearly, annually etc.

*4. **Reports***: Government offices, different ministries, census office, research institutes, technical institutes, NGOs etc. publish their reports to highlight the development related news. Several commissions setup by the Government time to time also submit their reports. Census organisation of India also prepares and submits reports after every ten years. Some evaluation agencies also submit reports. Some of these reports are published, while others remain unpublished but are maintained at the headquarters of the Institution concerned. The study of these reports provide the readers knowledge of development efforts made in different fields by different institutions. They also reveal the impact of development programmes as well as success and failures of development programmes. These reports also act as eye opener to politicians and bureaucrates who formulate development policies and implement them.

*5. **Articles***: A number of scholars dealing with policy formulation, implementation, monitoring and evaluation conduct study and publish articles on theoretical and practical aspects of development programmes. These articles are published in newspapers and research journals of national and international repute. They are helpful for policy makers, programme implementors and scholars interested in the study of development programmes.

*6. **Booklets***: A number of institutes, government offices, ministries, NGOs and scholars interested in studying development programmes get booklets published. These booklets contains development news of the concerned departments and institutions. The study of these booklets by scholars provide knowledge to understand theoretical and practical aspects of development programmes going on in our country.

*7. **Books***: A number of scholars of social science discipline like anthropology, sociology, history, economics and political science study development policies, plans, programmes etc. They also make evaluation study of the development programmes. They get their findings published with the help of publishers. The study of those books provide knowledge to us on different aspects of development policies, plans, programmes. They help us in formulating hypotheses and testing them in field for theory building.

8. Seminar/Workshop/Conference: A number of institutions get fund from the Government for holding seminar\workshop\conference on different aspects of development. Scholars from different parts of country participate in those seminars/workshops/conferences. They contribute theoretical and imperical papers on development policies, plans, programmes, monitoring and evaluation. The proceeding of the seminar is published and disseminated. They are very much helpful for the scholars interested in the study of development plans, policies, programmes, evaluation study etc.

9. Pictures and Cartoons: Artists and cartoonists publish their pictures and cartoons dealing with development, policies, programmes and their impact. They tell us their feeling through artforms and cartoon forms. They arouse the consciousness of the readers through their presentation.

10. Puppet-show: Puppet show is also a significant media of communication. Puppets are printed on screen or in art form. They sing song and perform dance. Puppet shows attract crowd in fair and weekly haat. They also attract people of the villages. Children like puppet show very much. If developmental news is communicated through puppet show and dance, it will be definitely beneficial for the beneficiaries.

On the basis of discussions on various types of media and their role in development, it can be said that media play significant role in development process, planning, policy formulation and impact result.

11

Human Rights and Development

Man is a developmental being. For the purpose of development he has made a number of inventions and still is engaged in making inventions. On the basis of inventions, undoubtedly he has made progress. But the pace of development and progress is not same for every ethnic group and regions of the world. As a result now the world is divided into developed, developing and underdeveloped. We also find such terminology as first world nations, second world nations and third world nations. In course of progress and development exploitation, oppression and deprivation of certain ethnic groups have also become a reality. The division of world and man in to rich and poor, developed and undeveloped and powerful and powerless, have given birth to 'have' and 'have not' exploiter and exploitee oppressor and oppressee, lord and serf, master and slave, etc. The existence of these categories clearly reveal that all human beings are not enjoying their human rights. The poor, indigenous people and weaker sections are not enjoying their human rights. They have been deprived and are still being deprived of their human rights. Considering the issue of Human Rights violation, the UN adopted the Universal Declaration of Human Rights on December 10, 1948. The UN proclaimed it as a common standard of achievement for all people and all nations. Without the enjoyment of Human Rights, development to all is not possible.

Human Rights are those rights of human beings which are essential for the development of all human beings and are expected to be enjoyed by all human beings of the world equally. Human Rights are divided into two broad categories:

I. Personal, civil and political rights
II. Economic social and cultural rights

Everyone has right to life, liberty and security of person. No one shall be held in slavery. Everyone is equal before law. No one shall be arbitrarily arrested. Everybody has freedom of movement. The right to work, the right to social security and the right of association have been recognised. Thus, Human Rights are the statement of moral standards which must be followed by all civilised nations.

International recognition of human rights includes the following:

- Right of political choice
- Freedom of association
- Freedom of expression
- Right to enjoy one's own culture
- Freedom from arbitrary arrest and detention
- Right to an adequate standard of living including health, housing and nutrition
- The right to work.

During the last 58 years the list of rights has been expanded to include new issues such as right to development, capital punishment, children in armed conflict, compensation of victim, disability, discrimination, based on HIV/AIDS, enforced or voluntary disappearances, environment, impurity, indigenous people, migrant workers, peacekeeping operations, sale of children, terrorism, human right defenders, war crimes and many more. But assuring human rights to all people have become still a challenge. The incidence of Human Right violation appears in all parts of the world in one way or the other. The survey conducted in 1999, by Gallap International revealed widespread dissatisfaction so far as the respect for human rights was concerned. Discrimination based on race and gender was common concerned for the people of different parts of the world.

The Building Blocks: The major advances in human rights and human development came after the horrors of the Second World War. The 1945 Charter of United Nations, followed by Universal Declaration of Human Rights in 1948 ushered in a new area of international commitment to Human Rights. Among the other things, they include the following:

I. Universality of Human Rights based on equality
II. Realisation of Human Rights as collective goal of humanity
III. Civil, political, economic, social and cultural rights for all people
IV. Creation of international system to promote the realisation of Human Rights with institutions to set standards, establish international laws and monitor performance but without powers of enforcement
V. The establishment of state's accountability for human rights obligation and commitments under international law.

Work on Human Rights remained continued. But being polarised by the cold war, the historic of human rights was often reduced to a weapon in official propaganda for geopolitical interests. The western countries emphasized upon the civil and political rights pointing their fingers towards socialist and developing countries. The socialist and developing countries emphasised upon the economic, social and cultural rights criticising the western countries for their failure to secure these rights for all citizens. In 1960, two separate conventions were held. One for civil and political rights and other for economic, social and cultural rights.

Principal Human Rights Instruments: International Bill of Rights is Principal Human Rights Instrument. International Bill of Rights consists of the Universal Declaration of Human Rights (UDHR), the International Covenant on Civil and Political Rights (ICCPR) and International Covenant on Economic, Social and Cultural Rights (ICESCR). The UDHR was a declaration adopted by General Assembly and hence requires no ratification. The ICCPR was ratified by 144 countries and the ICESCR was ratified by 142 countries.

Civil Rights: Civil Rights declared under UN Human Rights Declaration strongly condemn the existence of slavery and bondage of any kind. At the same time, it does not allow any kind of discrimination based on age, sex, race, caste, creed, ethnicity, region religion and language. The individuals belonging to ethnic groups of the world are entitled for following types of Civil Rights:

I. Equal right since birth
II. Equality before law
III. Equal opportunity of protection
IV. Equal opportunity of thinking and expression
V. Equal opportunity of practising own religion.

Social Rights: Under UN Declaration of Human Rights, the following Social Rights have been provided to the individuals of the world:

I. Every individual has right to educate himself/herself.
II. Every adult individual has right to enter into family life by arranging his/her own marriage.
III. Every adult male and female has right to lead married life.
IV. Family as fundamental unit of society must be protected by the Government of the concerned states and the society.

Economic Rights: The individuals of the world have been provided the following Economic Rights under UN Declaration of Human Rights:

I. Every individual has right to keep property
II. Every individual has right to increase his/her property.
III. Every individual has right of social security. Social security is such an arrangement in which the Government pays allowances to those who have very

little income or who do not possess any means of income

IV. Every individual has right to seek work

V. Every individual has right to get equal wages for equal work

VI. Every individual has right to work a specific period in a day and to get leisure for lunch .

Cultural Rights: The individuals of the world have been provided with the following Cultural Rights:

I. Every individual has rights to participate in cultural activities of the community to which they belong

II. Every individual has right to participate in arts of the community

III. Every individual has right to get benefit from cultural inventions and innovations.

Political Rights: In UN Declaration of Human Rights, the following Political Rights have been provided to the individuals of the world:

I. Every individual has right to participate in Government formation of the nation to which they belong.

II. Every individual has right to get appointment in Government job

III. The base of ruling in the country should be the will of the people

IV. Every individual has right to express opinion through election

V. Every individual has right of nationality.

International Convention on Elimination of All Forms of Racial Discrimination: It was adopted in 1965 and came in to force in 1969. It deals with a particular kind of discrimination based on race, colour, descent or national ethnic origin. It was ratified by 156 countries.

Convention on Elimination of All Forms of Racial Discrimination: It was adopted in 1979. It came into force in

1981. This was the first internationally and legally binding instrument to prohibit discrimination against women. It obligates Governments to take affirmative action for maintaining gender equality. It was ratified by 165 countries.

Convention against Torture, Cruel, Inhuman and Degrading Treatment or Punishment: It was adopted in 1984 and came in to force in 1989. The convention lays out steps to be taken by the Government to prevent torture, cruel, inhuman and degrading treatment or punishment. It was ratified by 119 countries.

Convention on Rights of Child: It was adopted in 1989. It came into force in 1991. This convention emphasises special attention on protecting and promoting the rights of children. It was ratified by 191 countries.

International Human Rights Committee: An International Committee on Human Rights have been setup. This committee has one chairman and four members by election. This committee looks after the protection and promotion of human rights and suggests steps to be taken for effective implementation of Human Rights. All member countries have voting rights.

Human Rights of Women: A woman is expected to enjoy the following Human Rights:

- A woman has right of equality since her birth
- She is equal before law
- She possesses right of freedom and protection
- She has freedom of thinking and expression
- She is free to practice her own religion
- She has right to participate in election and Government formation
- She has rights to educate herself
- She has right of nationality
- She has right to seek jobs in Government and private sectors
- She cannot be made slave and bonded
- She possesses freedom of association
- She possesses right to be free from arbitrary arrest and detention

- She has right to enjoy her own culture
- She has right of adequate standard of living including housing, nutrition, health and sanitation.
- She has right to work
- She has right to get equal wages for same kind of work
- She has right to possess, increase and inherit property
- She has right of social security
- She has right to work for a specified period in a day
- She has right to enjoy rest hours and holidays
- She has right to get payment daily or weekly or monthly
- She has right of maternity leave
- She has right to select life partner from her own choice.
- She has right to seek divorce
- She has right to participate in cultural activities of the group
- She has right to adopt arts of the community
- She has right to get benefit from scientific inventions and discoveries

Human Right of Children: The children are expected to enjoy the following Human Rights:

- Children have equal right to live since birth
- Children have right of equality before law
- Children have right of protection
- Children have right of thinking and expression
- Children have right of nationality
- Children have right to educate themselves
- Children have equal right of practising their own religion
- Children have right to enjoy ones own culture
- Children have right to be free from arbitrary arrest and detention
- Children have right of adequate standard of living including house, health, nutrition, sanitation and healthy environment
- Children have right to lead their childhood
- Children have right to remain free from slavery,

bondage, serfdom etc

- Children have right not to work as child labour
- Children have right to participate in cultural activities of the community
- Children have right to adopt the arts of the community
- Children have right to get benefit from scientific inventions

Human Rights of Labour: A labourer is expected to enjoy the following Human Rights:

I. Every labourer has right to lead his/her life in respectful manner

- Every labourer has right to seek work according to his/her own will and ability
- Every labourer has right to sell their labour in open market
- Every labourer has right to bargain his wages
- Every labourer has right to lead a free life
- Every labourer has right to remain free from contract, bondage, serfdom
- Labourer has right to accept or reject the work proposal
- Every labourer has right to work
- Every labourer has right of equal wages for same kind and hours of work
- Every labourer has right to work for a specified period in a day
- Every labourer has right to enjoy rest hours for lunch.
- Every labourer has right to avail leave
- Every labourer has right of protection from exploitation, oppression and deprivation
- Every labourer has right of association
- Every labourer has right of social security
- Every labourer has right to be free from arbitrary arrest and detention
- Labourer has right to enjoy one's own culture
- Labourer has right of opinion and expression
- Labourer has right to an adequate standard of living

including health, housing, food nutrition, sanitation, healthy environment etc

- Labourer has right to practice his/her own religion
- Labourer has right to participate in cultural activities of the community
- Labourer has right to adopt the arts of the community
- Labourer has right to enjoy the benefits of scientic inventions
- Labourer has right to lead a married life and family life
- Labourer has right to participate in election and Government formation
- Labourer has right of nationality
- Labourer has right to keep, increase and inherit property .

Human Right of Indigenous People: Indigenous people are expected to enjoy the following Human Rights:

- Indigenous people have right to lead their life in respectful manner
- Indigenous people have right to seek work
- Indigenous people have right to sell their labour in open market
- Indigenous people have right to bargain their labour
- Indigenous people have right to lead a free life
- Indigenous people have right to remain free from slavery, bondage, serfdom and any kind of contract
- Indigenous people have right to accept or reject the work proposal
- Indigenous people have right to get equal payment for same kind and hours of work
- Indigenous people have right to be free from arbitrary arrest
- Indigenous people have right of social security
- Indigenous people have right of survival
- Indigenous people have right to form association
- Indigenous people have full right of protection from exploitation, oppression deprivation and

discrimination
- Indigenous people have full right to enjoy their cultural herritage
- Indigenous people have right to maintain their relation
- Indigenous people have right to participate in cultural activities of the community
- Indigenous people have right to express their opinion
- Indigenous people have right to participate in election and Government formation
- Indigenous people have right of nationality and citizenship
- Indigenous people have right to adopt the arts of the community
- Indigenous people have right to enjoy the benefits of scientific inventions
- Indigenous people have right to keep, increase and in herit property
- Indigenous people have right to lead a married and family life
- Indigenous people have right to an adequate standard of living including health, housing, food , nutrition, sanitation and hygienic environment, drinking water etc

Violation of Human Rights in our Country: In our country the human rights of weaker sections such as Scheduled Castes and Scheduled Tribes, children, women and oldaged are subjected to violation.

Violation of human rights of the weaker sections are encouraged by rich, powerful industrialists and even Government officials. Violation of human rights take place in the following manners:

- Women are paid less than their male counterparts. This is also true for Government officials and contractors.
- The rich employ children for domestic and non-domestic works on very little wages or even on clothes and food.
- The rich employ aged people of the weaker section on a very nominal wages

- Women are sexually harassed at work sites
- Women are subjected to rape and gang rape when they go out for work
- Girls are prohibited to go to schools
- Girl children are mortgaged and sold like commodities
- Labourers are forced to work as bonded labour
- Migrant labourers have to live and work in unhygienic condition
- Labourers even do not get fresh drinking water at work site
- Housing conditions of the poor are unhealthy because they lack bath room, latrines, kitchens, drains etc
- The poor are bound to remain malnutrient
- The industrialists are making indigenous people up rooted from their land and culture
- In absence of proper rehabilitation, they are bound to become extinct
- Industrialists are creating imbalance in nature, man and spirit complex
- Industrial wastes throw in river are making the river water polluted. The poor people have to utilise those waste water of the river, which affect their health adversely
- Illeagal felling of trees of forests had made tribals un-employed as well as they have to bear cultural loss
- The people of weaker sections are subjected to exploitation, oppression and deprivation
- The people of weaker sections are prohibited to caste their votes.
- The booths of scheduled castes are looted and they are terrorised not to cast votes
- Caste riots, communal riots, lingual riots and regional riots are against human rights
- Murder, group murder and organised crimes are also against human rights

Agencies to Maintain Human Rights: In our country chief agency to maintain human rights is National Human Rights

Commission. Besides National Human Rights Commission, there is state level human rights commission, too. One can approach to these commissions if his/her human rights are violated by anyone.

Human Rights and Development: There is close relation between Human Rights and Development. For development, enjoyment of human rights is a must. Without enjoying human rights one cannot think of development. It is because violation of human rights is against the development. Human rights are violated forcefully under the compulsion of cultural traditions. Some cultural traditions make taboo on women, children and others. Those taboos go against the development and enjoyment of human rights. If girls are not sent to school, they are bond to remain illiterate. If women are kept only inside the house, they cannot participate in development process. If SCs are terrorised not to cast their votes, they do not participate in the democratic process of Government formation. If they are exploited, oppressed and deprived, they are bound to live like animal. Thus, for the development, encouragement of human rights is a must.

12

Panchayati Raj and Development

Panchayat is one of local self-governments in our country. Our country India is a modern state. As modern state its work is very extensive. It gets its work done through the Government. Our country has Central Government at the Centre and State Government at state level. Central Government performs works related to national importance. The responsibility of protecting the entire country lies at the shoulder of the Central Government. The State Governments maintain the administration of the states. In order to provide administration in the entire area of the state, a state is divided into commissionaries, districts, sub-divisions and thanas. There are Government officials at each level to look after the administration but besides administration villages and towns and cities have daily works at local level. Such as supply of electricity, drinking water and cleanliness of road, street, drains etc. It is not possible to perform these works by Central and the State Governments. These are local necessities and they have local importance. Therefore, local people only can solve these problems. The responsibility of local importance works lies at a third institution, which is termed as local self-government. Transaction of services related to the local needs of the people through local institution is known as local self-government. Local self-government provides right of administration to villagers, town dwellers, city dwellers and metropolis dwellers. In towns the institution of local self-government are municipalities, in cities there are municipal corporations and in metropolis there are mega municipal corporations, for cantonment, there are cantonment boards and for ports, there are port trust. For villages, the institutions related to the local self-government are Gram panchayat, panchayat samiti or Prakhand samiti and Zila Parishad.

Panchayati Raj and Constitution

Rashtra Pita Mahatma Gandhi was in favour of giving right of administration and development through local self-government, the panchayat. Pt. Jawaharlal Nehru and B.R. Ambedkar were also in favour of the governace of the villages through panchayats.

On 22 November, 1948, a motion was moved in parliament; after the Article 31, that the state shall take steps to organise village panchayats and endow them with such powers and authority as may be necessary for enabling them to function as units of self-government. The motion was unanimously adopted.

The involvement of people in the development through local self-government in villages was introduced in the First Five Years Plan (1951-56). Rajasthan was first state to introduce Panchayati Raj on 2 October, 1959 at Nagore. 5 days later Andhra Pradesh state introduced Panchayati Raj system. By mid-1960 Panchayati Raj was introduced across the country.

In the First Five Years Plan period attention was not paid towards the establishment of Panchayati Raj. Rather emphasis was given on community development project and National Extension Service Programmes. These programmes failed to bring out the participation of the people. This attracted the policy makers of Second Five Years Plan. In Second Five Years Plan emphasis was given on the establishment of the democratic institutions, i.e., Panchayat within the district where entire community could get involved in the development activities. For this purpose Balwantrai Mehta committee was appointed to study and report.

A number of committees were setup by the state Governments to assess the progress made by panchayats and to suggest measures for their improvement. For example, Purushottam Pai Committee (1964), Ramchandra Reddy Committee (1965), Narsimhan Committee in Andhra Pradesh,

Basappa Committee in Karnataka, Bongiwan Committee in Maharashtra, Vyas Committee in Rajasthan, Govind Sahay Committee in 1959 and Murti Committee in U.P.

The Department of Rural Development, government of India in 1978 set up Ashok Mehta Committee to assess and review the Panchayati Raj institutions in the country. Ashok Mehta Committee reported that Panchayati Raj was able to make average citizens more conscious of thier right politically. Administratively, Panchayati Raj institutions were able to bridge the gap between bureaucratic elites and the people socio-culturally. Panchayati Raj institutions generated a new leadership which was not only young in age, but modernistic and pro social change in outlook. From the point of view of development, Panchayati Raj institutions were able to help people cultivate a developmental psyche.

In Janata Party Government (1977), the Panchayati Raj got momentum. Ashok Mehta Committee was established. The committee in its reports had suggested for constitutionalising of panchayats. He had also made suggestion for a two tier system of panchayats, at District and Mandal levels. He had also suggested for the representation of SCs and STs as per their population in the election of the Panchayats.

From 1985 and onwards attempt to make Panchayats as institution of self-government was accelerated in between December 1987 and June 1988. Five workshops of District Magistrates were organised at different places in the country. Our Prime Minister had also participated in the workshop. These workshops recommended that a democratic framework at local level was indispensable for responsive administration. At the end of 1988, Panchayati Raj System in India was recommended for constitutional status.

On 15 May, 1989, Rajiv Gandhi Government introduced the 64th Constitution Ammendment Bill in Lok Sabha, but this Bill was defeated in Rajya Sabha.

Panchayati Raj (Amendment) Act, 1993

Again on 24 April, 1993 the Nars Simha Rao Government introduced 73rd Constitution Amendment Bill in 1992. The Bill was passed in both Houses. It became Act on 24 April, 1993.

This is also know as Panchayati Raj Act. Some of salient features of 73rd Ammendment Act, 1993 are as follows:

1. The Act stresses at the formation of Gram Sabha.
2. The Act emphasises upon uniform three tier system of Panchayati Raj at village, Block and District level with exemption for intermediate level in states with population less than two millions.
3. It emphasis upon direct election to all seats for all members at all levels.
4. The Act has decided 21 years as the minimum age for membership as well as for chairpersons.
5. The Act has provided reservation on rational basis for SCs and STs in proportion to their population for membership as well as chair person of the panchayats.
6. The Act has provided reservation of not less then one-third of seats for women.
7. According to this Act, the Term of a Panchayat will be of Five years.

Now-a-days, Panchayati Raj system has been established in our country with constitutional approval.

Panchayati Raj System is a three-tier system. At village level, there is Gram Panchayat, at the Block Level, there is Panchayat Samiti or the Prakhand samiti. At the district level, there is Zila parishad.

Description about Gram Panchayat, Panchayat Samiti and Zila Parishad has been presented under the chapter Governmental Agencies of Development. Therefore, it is not necessary to present them again.

Panchayati Raj Scheduled Area Extension, Act 1996

This Act dealing with Scheduled Area has come into force in 1996. This Act lays emphasis upon the reservation in panchayats in Scheduled Areas. The main features of Panchayati Raj Scheduled Areas Extension Act are as follows:

1. It lays emphasis upon reservation of seats in Panchayats of Scheduled Area for Scheduled Tribes and Scheduled Castes in proportion to their population

2. Reservation of seats for Scheduled Tribes should not be less than half of the total seats in Panchayats.
3. The posts of chair person at all levels such as Gram sabha, Panchayat samiti and Zila Parishad will be reserved for Scheduled Tribes in the Scheduled Area.
4. The State Government may nominate a person of such Scheduled Tribe whose representation is lacking at panchayat or district level. But nomination of such persons should not be more than one-tenth of the total elected members of that panchayat.

Extra Power in Scheduled Area

The Gram Panchayats have been provided extra power in Scheduled Areas. Extra powers provided to panchayats of Scheduled Areas are as follows:

1. Maintenance of cultural identity and traditional resources of the communities.
2. Maintenance of village Haats and cattle fairs.
3. Organisation of more than one Gram sabha in a village which will function in traditional manner.
4. Chairman of Gram Sabha of Scheduled Areas will not be mukhia, upamukhia or member, but village head, munda, manjhi, manki etc.

Powers Given to Gram Sabha in Scheduled Area Extension Act, 1996

As per the Act of 1996, the Gram Sabha of Scheduled Areas have been given the following powers:

1. Right of management of natural resources like land, water, forest etc.
2. Management of wells, ponds minor irrigation, etc.
3. Management of mineral resources of the villages.
4. Control on sale and purchase of intoxicating materials.
5. Collection, storage, processing and marketing of minor forest produce.

6. Restoration of tribal lands and development of lands distributed among landless labourers.
7. Management of village haat and cattle fair
8. Prohibition on money lending, encouragement, towards profit and organisation of Self Help Group.

Impact of Panchayati Raj on Tribal, and Rural Development

The impact of Panchayati Raj on Tribal and Rural Development can be discussed under two broad headings: (i) Positive Impact and (ii) Negative Impact.

Positive Impact: Panchayati Raj institutions have brought the following positive impact on the tribal and rural development in our country:

1. Tribal and rural populations have become conscious of their political rights in the election of Panchayats and Zila parishad.
2. Panchayati Raj institutions have provided tribal and rural people with occasion to put their problems in Gram Sabha.
3. Panchayati Raj institutions have provided occasion to tribal and rural people not only to exercise only voting power, but also to fight elections for members or chairpersons of Panchayat, panchayat samiti and Zila Parishad.
4. Tribal and rural people have become aware of rural development programmes such as road, street, drinking water, electricity, school, hospitals etc.
5. Panchayati Raj institutions have encouraged the tribal and rural people to participate in rural and tribal development programmes.
6. Tribal and rural people have been able to recognise their political power on the basis of their numerical strength.
7. Tribal and rural people have been able to stand against exploitation, oppression and deprivation.
8. Tribal and rural people have got occasion to develop their communities and villages.
9. Panchayati Raj institutions have lessened the distance

between rich and poor and high castes and low castes.

10. Tribal people and Rural People have come in contact with NGOs which are working for their welfare and development.
11. Panchayati Raj institutions have made SCs, STs and women empowered.
12. Panchayati Raj institutions have created awareness towards education.
13. Panchayati Raj institutions have made people aware of drinking water, health, sanitation and environment conservation.
14. Agriculture and minor irrigation schemes have got successful implementation in tribal and rural areas.
15. New technologies have reached in tribal and rural villages.
16. Tribal and rural people have been integrated to development programmes.
17. Panchayati Raj institutions have made weaker sections aware of power and right.
18. Panchayati Raj institutions have increased the participation of tribal women and rural women in development process. They are holding the posts of mukhia, pramukh, upapramukha and chairman Zila Parishad.

Negative Impact: Panchayati Raj System has influenced the tribal and rural development negatively in the following ways:

1. It has brought negative impact on traditional political system like village head, council of elders, village Panchayat, inter village Panchayat etc.
2. It has opened the way of thana, police, court and lawyer among tribal and rural people.
3. It has encouraged factionalism and feuds in tribal and rural communities.
4. Bloodshed takes place at the time of election of mukhia and Zila Parishad.
5. Incidences of booth capturing and murders have appeared in the election of the Panchayats.
6. It has created a gap between tribals and non-tribals in

tribal dominated areas.

7. Caste and tribe conflicts have come into existence.
8. Tribals have launched agitations to keep non-tribals out of their territory.
9. Tribal identity is being questioned.
10. Ethnic movements have come into existence.
11. Concepts of sub-nationalism and regionalism have emerged.
12. Demand for separate state is increasing.

There are 2,34,078 Panchayati Raj institutions in our country. A detail description about them is as follows:

Takes of Panchayat	*No. of Panchayati*
1. Gram Panchayats	2, 27, 998 at village level
2. Panchayat Samitis	5, 906 at block level
3. Zila Parishad	427 at district level
Total	2,34,078

The total elected members of the Panchayati raj institutions are 28,79,023 and development functionaries are 10,27,167.

Type of Panchayat	*Elected Members*	*Development functionaries*
1. Gram Panchayat	27,15,090	9,04,908
2. Panchayat Samitis	1, 48, 555	1,06,513
3. Zila Parishad	15, 378	12,144
Total	28,79,023	10,23,167

Distribution of elected members in to SCs, STs and women are as follows:

	SCs	*STs*	*Women*
1. Gram Panchayat	3,08,290	11,99,401	5,77,270
2. Panchayat samitis	18,797	8,479	37,611
3. Zila Parishad	1,794	1,165	2,516

13

Development Strategy: Five Year Plans

After independence, our Government adopted a five years plan policy as development strategy. From the time of implementation of First Five Years Plan till now Ten Five Years Plan periods have passed. Now we are in 11th Five Years Plan. In these ten Five years plan periods, a number of development programmes were formulated and implemented. At the end of each five years plan, development models and programmes were evaluated by the expert committee. Some programmes were continued and others were dropped on the recommendation of the committee setup by the Government. Now it will be the fitness of the thing to provide description of policies and development programmes of each plan period one be one.

First Five Years Plan (1951-1956)

The First Five Years Plan started in April 1951 and continued till 1956. The main aim of this plan was to establish a sound economy of the country so that several big programmes could be formulated and implemented in future. The total expenditure in this plan was Rs. 2378 crore but actual expenditure could be met only Rs.1960 crore. Out of Rs. 1960 crore, Rs. 174.7 crore was for the welfare of the tribals. The expenditure in this plan was as follows:

Sl. No.	Item	Expenditure in Rs. crore	Percentage of total Expenditure
1.	Agriculture and Allied sector	290	14.8
2.	Irrigation and Flood control	434	22.2
3.	Electricity (Power)	149	07.6
4.	Village and Small Industry	42	02.1
5.	Industry and Minerals	55	02.8
6.	Transport and Communication	518	26.4
7.	Others	472	24.1
	(a) Education and scientific Research	149	07.6
	(b) Health and Family Planning	98	5.0
	Total	1960	100.0

From the above account it is clear that in First Five Years Plan emphasis was given on agriculture, community development, irrigation and electricity. Transportation and communication had also received attention. 24.1 per cent expenditure was made on social service and others. Sine the first plan did not envisage any large scale progrmmes of industrialisation, the sector of industries and minerals received only 2-8 per cent of total allocation.

Second Five Years Plan. (1956-1961)

The second Five Years Plan was started in April 1956 and ended in March 1961. It was based on Mahalanobis model. This model emphasised upon the adoption of large scale industrialisation programmes in the country. The development out lays under the second Five year plan was as below:

Sl.	Head of Development	Second Plan 1956-61 Actual Rs. crore	Percentage to Total
1.	Agriculture and Allied Sector	549	11.7
2.	Irrigation and Flood Control	430	09.2
3.	Power	452	09.7
4.	Village and Small Industry	187	04.0
5.	Industry and Minirals	938	20.1
6.	Transport and Communications	1261	27.0
7.	Others of which	855	18.3
	(a) Education and Scientific Research	273	5.8
	(b) Health and Family Planning	216	4.6
	Total	4672	100.0

From the above account of the second Five years Plan, it is clear that the share of industry and minerals was raised from 2.8 per cent in First Five years plan to 20.1 per cent in second Five years Plan. The allocation to agriculture and allied sector declined from 14.8 percent to 11.7 percent under the second plan. However, in absolute term it was doubled from 290 crore to 549 crore. Transport and communication sector had attained allocation of 27.0 percent. Power sector had received allocation of 9.7 percent. The resource allocated to village and small industries increased four and half time as compared to first plan, but share in total outlay was a mere 4.0 percent.

Third Five Years Plan. (1961-1966)

The Third Five Years Plan began in April 1961 and ended in March 1966. The Third Plan had following objectives:

1. To achieve 30 per cent increase in National income.
2. To attain self-sufficiency in food grains
3. Establishment of steel, electricity, oil and fuel industries
4. Utilisation of man power and increase in means of employment
5. To end the disparity in distribution of wealth and income.

The Development outlays of Third Five Years Plan was as follow:

Sl. No.	*Development Item*	*Expenditure in Third Plan (1961-66) Rs. crore*	*Percentage to Total Expenditure*
1.	Agriculture and Allied Sector	1089	12.7
2.	Irrigation and Flood Control	665	7.8
3.	Power (Energy)	1252	14.6
4.	Village and Small Industry	241	2.8
5.	Industry and Minerals	1726	20.1
6.	Transport and Communication	2112	24.6
7.	Others of which	1492	17.4
	(a) Education and Scientific Research	660	7.7
	(b) Heath and Family Planning	251	2.9
	Total	8577	100.0

From the above table, it is clear that the third plan was double in size of the second plan. The total expenditure rose from Rs. 4672 crore in first plan to Rs. 8577 crore in third plan. The Third Plan continued the effort initiated in second plan and envisaged no change in priority. The outlay on agriculture and allied sector was doubled from Rs. 549 crore in second plan to Rs. 1089 crore in third plan. Similarly, the outlay on industries and minerals was increased from Rs. 938 crore in second plan to Rs. 1726 crore in Third Plan. As in First and second plans, highest allocation in third plan also went to transport and communication sector. It was due to expansion of industrialisation process. In this plan, there was an outlay of Rs. 2, 112 crore for transportation and communication sector. Of this, 890 Rs. Crore was earmarked for the development of railways.

The Period of Annual Plan

The Third Five Years Plan was followed by period of a three Annual plans were as fourth plan was postponed by three years. The out lays of Annual plans were as follows:

Sl. No.	*Head of Development*	*Expenditure in 1966-69 Rs. crore*	*Percentage to Total Expenditure*
1.	Agriculture and Allied Sector	1107	16.7
2.	Irrigation and Flood Control	471	7.1
3.	Power (Energy)	1212	18.3
4.	Village and Small Industries	126	1.9
5.	Industry and Minirals	1510	22.8
6.	Transport and Communication	1222	18.5
7.	Others of which	977	14.7
	(a) Education and Scientific Research	354	5.3
	(b) Health and Family Planning	210	3.2
	Total	6625	100.0

From the above table it is clear that the total actual expenditure under these three annual plans was 6625 crore of which 22.8 per cent to industries and minerals, 18.3 per cent to power, 18.5 per cent to transport and communication and

16.7 per cent to agriculture. A significant change that one can observe was accorded to development of power from 7.6 per cent in first plan, 9.7 per cent in Second plan, 14.6 per cent in third plan and 18.3 percent in three annual plans. This had become necessary due to power shortage prevailing in the country which had affected industrial production adversely.

Fourth Five Years Plan (1969-74)

The actual expenditure under the fourth plan was Rs. 15,779 crore as against the plan outlays of Rs. 15902 crore. The development out lay under Fourth Five Years Plan was as follows:

Sl. No.	*Head of Development*	*Outlay in 1969-74 Rs. crore*	*Percentage to Total Expenditure*
1.	Agriculture and Allied Sector	2320	14.7
2.	Irrigation and Flood Control	1354	8.6
3.	Power Energy	2932	18.6
4.	Village and Small Industries	243	1.5
5.	Industry and Minerals	2864	18.2
6.	Transport and Communication	3080	19.5
7.	Others	2986	18.9
	(a) Education and Scientific Research	905	5.7
	(b) Health and Family Planning	613	3.9
	Total	15779	100.0

The above table reveals that the highest allocation was kept for transport and communication as in earlier plans, i.e. 19.5 per cent. In this plan power sector had attained an outlay of 18.6 per cent. Industries Sector received 18.2 per cent of the total outlay of this plan. Village and small industries got only 1.5 per cent share. Agriculture and Allied sector got only 14.7 per cent. Irrigation and Flood control got 8.6 per cent share.

Fifth Five Years Plan (1974-79)

The total outlay was Rs. 39426 crore. The sectoral distribution of the outlays in Fifth Five Year Plan was as follows:

Fifth plan proposed a public sector outlay of Rs. 37,250 crore whereas the actual expenditure was Rs. 39,426 crore. The highest priority was accorded to industry in this plan. Its share was 22.8 per cent. Power sector got 18.8 per cent share. Agriculture and Allied sector had attained 12.3 per cent share. The share of irrigation and flood control was 9.8 per cent and village and small industries 1.5 per cent. Transport and communication had attained 17.4 per cent. Education and scientific research had attained 4.4 per cent share, while the share of Health and Family Planning was 3.9 per cent.

Sl. No.	*Head of Development*	*Outlay in 1969-74 Rs. crore*	*Percentage to Total Expenditure*
1.	Agriculture and Allied sector	4865	12.3
2.	Irrigation and Flood Control	3877	9.8
3.	Power	7400	18.8
4.	Village and Small Industries	592	1.5
5.	Industry and Minerals	8989	22.8
6.	Transport and Communication	6870	17.4
7.	Others	6833	17.4
	(a) Education and Scientific Research	1710	4.4
	(b) Health and Family Planning	1253	3.2
	Total	39426	100.0

For the first time in Indian planning history, the Approach paper of Fifth Five Year Plan talked of a Redistributive Programme aiming at bringing down the per capita consumption level of the richest 30 per cent of population so that the lowest 30 per cent of the population could be guaranteed a minimum level of subsistence. The following objectives were laid down to achieve this objective:

(i) By the final year of Fifth plan, poverty should be removed in the sense that practically no body should have a consumption level below a certain minimum.
(ii) Inequality of distribution of consumption should be reduced by a certain stipulated degree.
(iii) Aggregate consumption should be set for final year.
(iv) Other macro-economic targets of final year plan.
(v) Production structure at disaggregated level.

In agricultural sector, the thrust was on land development and improved agricultural, horticultural and live stock practices. Special programmes like Drought-prone Area programmes, small farmers and labourers development programme etc. also received due emphasis. In irrigation and power sectors the stress was towards completion of spill over projects and better utilisation of potential already available. In mining and manufacturing sector, high priority was accorded to achievement of increased production in basic lines particularly steel, non-ferrous metals, fertilisers, coal, machine building and mineral oils. Emphasis was made on increasing the output of mass consumption goods. Special emphasis was placed on raising the condition of weaker sections. For this purpose minimum need programme (MNP) was initiated. This included a wide spectrum of programmes covering elementary education, drinking water, medical care, in rural areas. Nutrition, house sites for landless labour, rural roads, rural electrification and slum clearence. The total outlay provided for this programme under Fifth plan was Rs. 3,312 crore, whereas the actual expenditure was Rs. 1578 crore.

Sixth Five Years Plan. (1980-85)

Fifth Five Years P lan period was 1974-79 and the Sixth Five Years Plan period was 1980-85. Thus, there was a gap year of 1979-80. In this gap year, an annual plan was operating. The development outlay under sixth plan period was as follows:

Sl. No.	*Head of Development*	*Plan Outlay (1980-85) Rs. in crore*	*Percentage to Total Expenditure*
1.	Agriculture and Allied Sectors	15201	13.9
2.	Irrigation and Flood Control Sectors	10930	10.0
3.	Power (Energy)	30751	28.1
4.	Village and Small Industries	1945	1.8
5.	Industry and Minerals	15002	13.7
6.	Transport and Communication	17673	16.2
7.	Others of which	17790	16.3
	(a) Education and Scientific Research	3997	3.6
	(b) Health and Family Planning	3442	3.1
	Total	109292	100.0

From the above table it is clear that the highest priority in this plan went to energy sector amounting to 28.1 per cent. Next to energy sector, second highest allocation was made to transport and communication. This sector had received 16.2 per cent to the total outlay. Industries and Minerals sector received 13.7 per cent share in plan outlay. Agriculture and allied sector received 13.9 per cent share. Village and small industries got 1.8 per cent share in plan outlay. Education and scientific Research got 3.6 per cent share, while Health and Family Planning got 3.1 per cent share in sixth plan outlays.

The sixth plan was formulated against the background of a perspective covering a period of 15 years from 1980-81 to 1994-95. This development perspective visualised accelerated progress towards the removal of poverty, generation of gainful employment, and technological and economic self-reliance.

Seventh Five Years Plan. (1985-1990)

The sectoral allocation of public sector outlay in Seventh Five Years Plan was as follows:

Sl. No.	*Head of Development*	*Plant outlay in Rs. crore*	*Percentage Total*	*Actual Expen-diture*	*Percentage*
1.	Agriculture	10, 574	5.9	12793	5.8
2.	Rural Development	9074	5.0	15247	7.0
3.	Special Area Programme	3145	1.7	3470	1.6
4.	Irrigation and Flood Control	16979	9.4	16590	7.6
5.	Energy (Power)	54821	30.5	61689	28.2
6.	Industry and Minerals	22461	12.3	29220	13.4
7.	Transport	22971	12.8	29548	13.5
8.	Communication, Information Broadcasting	6472	3.6	8426	3.9
9.	Science and Technology	2466	1.4	3024	1.4
10.	Social Services	29350	16.3	34960	15.9
11.	Others	1687	0.9	3763	1.7
	Total	1,80,000	100.0	2,18,730	100.0

The sectorial allocation of public sector outlay of Seventh Five Years Plan reveals that the largest share in allocation went to energy sector. This sector got 30.5 per cent share in outlay. Agriculture, Rural Development, special Area programme and irrigation received together a share of 22.0 per cent in the plan outlay Industry and mineral sector received 12.3 per cent share. Transport sector got 12.8 per cent share. Communication, Information and Broadcasting got 3.6 per cent share. Science and technology received 1.4 per cent share. Social services received 16.3 per cent share of plan outlay thus, in seventh plan period attempt was made to ensure balance among infrastructure sectors, other production sectors and Human Resource Development Sector.

Eight Five Year Plan (1992-1997)

The second plan completed its terms on March 31, 1990. The Eighth plan should have logically commenced from April 1, 1990. But this schedule could not be maintained due to political uncertainty at the centre. For the year 1990-91 and 1991-92, annual plans were implemented. Thus, the Eighth Five Years Plan started in April 1992 and continued till March 1997. The sectoral allocation of public sector outlay of the Eighth plan was as follows:

Sl. No.	*Head of Development*	*Plan outlay (1992-97) in Rs. crore*	*Percentage of Total*
1.	Agriculture and allied Activities	22467	5.2
2.	Rural Development	34425	7.9
3.	Special Area Programmes	6750	1.6
4.	Irrigation and Flood Control	32525	7.5
5.	Energy	111561	26.6
6.	Industry and Minerals	46922	10.8
7.	Transport	55926	12.9
8.	Communication	25110	5.8
9.	Science, Technology and Environment	9042	2.1
10.	Social Services	79012	18.2
11.	Others	6360	1.4
	Total	430100	100.0

The above table reveals that the highest allocation of resources in Eighth plan went to the energy sector. This sector was allocated 26.6 per cent of resources. Agriculture, Rural development, special Area Programmes and irrigation together were allocated 22.2 per cent of the resources. Industrial and mineral sector had received only 10.8 per cent of the plan outlay. Transport sector was allocated 12.9 percent and communication was given 5.8 per cent share. Science, technology and Environment received 2.1 per cent and social services were allocated 18.2 percent.

The main objectives of the Eighth plan was as follows:

1. Generation of adequate employment to achieve near full employment level by the turn of the century.
2. Containment of population growth through active people's cooperation and an effective schemes of incentives and disincentives.
3. Universalisation of elementary education and complete eradication of illiteracy among the people in the age group of 15 to 35 years.
4. Provision of safe drinking water and primary health care facilities including immunisation accessible to all villages and entire population and complete elimination of scavenging.
5. Growth and diversification of agriculture to achieve self sufficiency and generate surplus for experts.
6. Strengthening the infrastructure in order to support the growth process on a sustainable basis.

The above list of objectives reveals clearly that the main concern of Eighth plan was with human development. The objective listed to achieve the goal were employment generation, population control, literacy, education, health facilities, drinking water, adequate food and basic infrastructure. It is surprising to note that economic growth did not appear in the list of objectives stated in Eighth plan. However the target growth rate was kept 5.6 per cent per annum. This growth rate was higher than trend growth rate achieved by Indian economy in entire planning period. Thus, economic growth continued to occupy significant place in the over all scheme of things in Eighth Five Year Plan.

Ninth Five Years Plan. (1997-2002)

The Ninth Five Years Plan officially commenced on April 1, 1997 and continued till March 2002. However, the Ninth Five Year Plan document was released by the Planning Commission only in March 1999. The delay was due to causal approach now adopted towards the entire planning exercise. The process of economic liberalisation unleased in 1991 and persued vigorously by successive Governments. We now moved from an era of dissipated development strategy. In this strategy we adopted declaration of intentions in place of objectives and projection in place of targets.

According to the Planning Commission, the focus of Ninth plan can be described as growth with social Justice and Equity. The specific objectives listed in this plan were as follows:

1. Priority to agriculture and rural development with a view to generating adequate productive employment and eradication of poverty.
2. Accelerating the growth rate of economy with stable prices.
3. Ensuring food and nutritional security for all, particularly the vulnerable section of the society.
4. Providing the basic minimum services of safe drinking water, primary health care facilities, universal primary education, shelter and connectivity to all in a time bound manner.
5. Containing the growth rate of population.
6. Ensuring environmental sustainability of development process through social mobilisation and participation of people at all levels.
7. Empowerment of women and socially disadvantaged groups such as Scheduled Castes, Scheduled Tribes, Other Backward Classes and minorities as agents of socio economic change and development.

8. Promoting and developing people's participatory institutions like Panchayati Raj institutions, cooperatives and self Help Groups.
9. Streng thening Efforts to build selfsreliance.

Public sector outlay by Major Heads of Development in Ninth plan was as follows:

Sl. No.	*Development Head*	*Outlay Rs. crore*	*Percentage to Total*
1.	Agriculture and Allied Activities	37,546	4.4
2.	Irrigation and Flood Control	55, 598	6.5
3.	Rural Development	73,439	8.5
4.	Special Area Programme	3,649	0.4
5.	Energy	2,15,545	25.1
6.	Industry and Minerals	69,972	8.1
7.	Transport	1,21,324	14.1
8.	Communication and Information Technology	47,616	5.0
9.	Science, Technology and Environment	25,529	3.0
10.	General Economic Services	15,038	1.8
11.	General Services	11,940	1.4
12.	Social Services	1,82,205	21.2
	Total	8,59,200	100.0

From the above table, it is clear that 25.1 per cent of total public sector outlay was earmarked for energy sector. Energy is one of the most important inputs required for economic and social development. In fact, inadequate development of energy sector can prove to be most important stumbling block in the industrial development of a country. Transport, communication and information technology is another sector which provides necessary infrastructure for economic and industrial development. This sector was allocated 19.6 per cent of total public sector outlay. Social services, include education, medical, public health, family welfare, housing, urban development etc. and they were jointly allocated 21.2 per cent of public sector outlay in Ninth Plan. Industry and mineral sector had received 8.1 per cent of total outlay. Agriculture, irrigation, rural development and special area programme had received jointly 19.4 percent of the total outlay. Science, techolology and environment got 3.0 percent allocation out of total outlay.

Tenth Five year plan (2002-2007)

The Tenth Five Years Plan was started on 1st April, 2002 and continued to exist till 31st March, 2007. Tenth plan invisages an annual average growth rate of 8 percent. Recognising that economic growth cannot be the only objective of national plan, the tenth plan has set monitorable targets for few key indicators of human development in addition to 8 per cent growth rate. These include reduction of poverty ratio by 5 percentage points by 2007; providing gainful employment to all.

The total outlay envisaged for the Tenth plan is Rs. 19,68,815 crore, comprising of an outlay of Rs. 7,06,000 crore for the central plan, Rs. 5,88,325 crore for the state plan and Rs. 6, 74,490 for the public sector enterprises. The budgetary support for this outlay is envisaged at Rs. 9,94,060 crore.

Macro parameters of 10th plan were as follows:

1.	Domestic saving Rate	26.84
2.	Current Account Deficit	1.57
3.	Investment rate	28.48
4.	International capital output ration	3.58
5.	GDP	7.93

14

Industrial Development, Displacement and Rehabilitation

From the Second Five Year Plan our Government began to pay much attention towards mega industrial projects. Following Mahalanobis model of development strategy, our Late Prime Minister Pt. Jawaharlal Nehru had played significant role in the development of mega industrial projects in naturally resourceful areas of our country. That is why the development of mega industrial complexes in our country is also known as Nehru model of development. The basic idea behind Mahalanobis Model or Nehru Model of mega industrial development was development of area and the people through mega industrialisation process and to bring the country under developed economy.

Industrial Development in First Plan

On the eve of the First plan, the industrial development in India was confined largely to the consumer goods sector, the important industries being cotton textiles, sugar, salt, soap, leather goods and paper. This feature of industrial development in the First plan exhibited the feature of undeveloped economy. Industries manufacturing intermediate products like coal, cement, steel, power, alcohol, non-ferrous metals, chemicals etc. were also exhibited but their production was small because productive capacity was considerably below the requirements except cement. On the whole consumer goods industries were well established, but producer goods industries lagged behind. Thus the first plan did not envisage any large scale programme of industrialisation. That is why only 2.8 per cent of the total plan outlay was allocated to industry and minerals. However, the first plan gave a practical shape to the concept of mixed economy in which public and private sectors grew in complementary manner. The public sector industries setup in this plan were Hindustan shipyard, Hindustan Machine Tools, Hindustan Antibiotics, Integral Coach Factory, UP Government cement factory etc. Private sector industries were also set up but in consumer goods sector like sugar, cotton textiles, sewing machine, automobiles, bicycles etc.

Industrial Development in Second Plan

The second plan accorded top priority to programmes of industrialisation. As a result 20.1 per cent of the total outlay of the plan was allocated to the head industry and minerals. In this plan substantial expansion in iron and steel, non-ferrous metals, coal, cement, heavy chemicals, and other industries of basic importance were given top priority in the industrial sector. Three steel plants of one million tonnes ingot capacity each were set up in the public sector at Bhilai, Rourkela and Durgapur besides expansion and modernisation programmes undertaken in the private sector.

Industrial Development in Third Plan

The Third Five Years Plan also stressed upon the establishment of basic capital and producer goods industries. In this plan special emphasis was placed upon machine building programmes so that the growth of economy in subsequent plan could become self-sustaining. Programmes of Heavy Engineering, machine building industries, castings and forgings alloy tool, special steel, iron and steel, ferro-alloys, step to increase fertilisers output and petroleum products were taken up. Efforts were also made to increase the production of major basic raw materials and producer goods like aluminium, mineral oils, dissolving pulps, basic organic and in organic chemicals and products of petro-chemical origin.

In the second and the third plans, great emphasis was placed on capital goods industries and basic industries. Most of these industries were setup in public sector. As a result the size of public sector grew rapidly. The growth of consumer goods industries was left to the private sector. The same structure of industrial development was nurtured and promoted in the fourth and fifth plan, too.

Industrial Development in Fourth Plan

In Fourth plan, industries and minerals were allocated 18.2 per cent of the total outlay.

Industrial Development in Fifth Plan

The expenditure on industry and minerals was increased to 22.8 per cent in Fifth plan.

Industrial Development in Sixth Plan

In an industrial development review over the thirty years of planning, the sixth plan observed that the industrial production had increased by about 5 times during 30 years of planning in our country. There was quantitative increase in output. The industrial structure was widely diversified covering broadly the entire range of consumer, intermediate and capital goods. However, the structure of industrial development was not sufficiently guided by cost consideration. There was prevalence of high cost industrial structure. The public sector had failed to generate enough resources. The problem of regional disparity in industrial development was as serious as ever. The new industries had tended to graviate towards centre and the backward area had remained untouched by the planning.

In the light of all above observations, the sixth plan emphasised upon the optimum utilisation of existing capacities and improvement of productivity, enhancement of manufacturing capacity, special attention to capital good industries, electronic industries, improvement in energy efficiency dispersal of industries, etc. In the sixth plan, the industry and mineral sector had received 13.7 per cent allocation share of the total outlay.

Industrial Development in Seventh Plan

The main objectives for industrial development in the seventh plan were as follows:

(i) To ensure supply of wage goods and consumer articles of mass consumption at reasonable prices and acceptable quality.
(ii) To maximise the utilisation of existing facilities through restructuring, improved productivity and upgradation of technology.
(iii) To concentrate on development of industries with large domestic market and export potential to emerge as world leader in them.
(iv) To usher in sunrise industries with high growth potential and relevance to our need.
(v) To evolve an integrated policy towards self-reliance in strategic fields and opening up of avenues for employment of skilled and trained man power. The actual expenditure on industries and minerals in the seventh plan was 11.9 per cent of the total expenditure.

Industrial Development in Eighth Plan

So far as the Eighth plan is concerned, overall outlay for industrial and mineral programmes was 9.3 per cent of the total outlay. This reduced allocation to industry and minerals is in the line with liberalisation policy announced in the new Industrial policy of 1981. According to this New Industrial policy private sector is now destined to play significant role in industrial activities especially in those fields where security and strategic or social considerations are not important.

Industrial Development in Ninth Plan

The Ninth plan outlay for industry and minerals was kept as 7.6 per cent of the total plan outlay. The Ninth plan envisaged growth rate as 8.2 per cent per annum. Main objectives to achieve this growth rate were as follows:

(i) Ensuring adequate availability and requisite quality of infrastructure.
(ii) Adoption of special measures to promote development of industries in backward area.
(iii) Introducing a special package for development of North Eastern States.
(iv) Reviewing the working of Board of Industrial and Financial Reconstruction (BIR).
(v) Initiating steps to close down potentially unrevivable public sector units.
(vi) Promoting production and productivity in small scale industries through technological upgradation and ensuring adequate credit to sector.
(vii) Adoption of a cluster approach in the unorganised sector, with provision of training, upgradation of skill and improvement in tool kits, equipments and production techniques.

Major Iron and Steel Industries

The earliest attempt to manufacture iron and steel by modern methods was made in the country at Barakar in 1875 for the production of pig iron. This was taken over by the Bengal Iron Company in 1889. The first attempt to produce iron and steel in our country at large scale was made when Tata Iron & Steel Company (TISCO) was set up at Jamshedpur in 1907. Indian Iron and Steel Company (IISCO) was setup at Burnpur in 1919. The first unit in public sector was set up at Bhadravati and started functioning in 1923. It is now known as Visvesvaraya Iron and Steel Works Ltd. This was the situation in pre-independence period.

After independence, special attention was paid to the development of iron and steel industries. The second plan gave top priority to the development of iron and steel industry. Three steel plants of one million tonnes ingot capacity each were established at Durgapur (West Bengal), Bhilai (Madhya

Pradesh) and Rourkela (Orissa). Expansion programme to double the capacity of two private sectors, namely TISCO and IISCO to 2 million tonnes and 1 million tonnes respectively were also taken into hand. Heavy Engineering Corporation was also established at Hatia. Ranchi in the second plan period. The third plan placed emphasis on expansion of these steel plants and setting up of a new steel plant at Bokaro. In fourth plan, three new steel plants were set up at Salem in Tamil Nadu, Vijayanagar in Karnataka and Visakhapatnam in Andhra Pradesh. The Bokaro steel plant was commissioned on February 26, 1978. Its starting ingot capacity 8.9 million tonnes was increased to 11.6 million tonnes on March 31,1980. The Gevernment took over the management of IISCO in 1972 and acquired its ownership in 1976 to improve its working.

Prior to 1973, the plants at Bhilai, Rourkela and Durgapur were owned and managed by Hindustan Steel Limited (HSL) and the Bokaro steel plant by Bokaro Steel Limited (BSL). In 1973, the Government set up Steel Authority of India Ltd. (SAIL). HSL and BSL became wholly owned subsidiaries of SAIL. The management of IISCO is also under SAIL. Visvesvaraya Iron and Steel Ltd was taken over by SAIL in August 1989. Thus, SAIL is now integrated steel company. Visakhapatnam Steel Plant of Rashtriya Ispat Nigam Ltd. was commissioned in July 1992. It is the best laid out steel plant in the country with a capacity of three million tonnes. In the private sector TISCO situated at Jamshedpur is the first integrated steel plant. Other players in private sectors are Essar, Mukand (biggest miniplant in the country), Liods, Jindal, Nippon Denro Ispat Ltd. Mahindra Ugine Steel Company Ltd., FACOR, Mardia Steel Ltd. India is now ninth largest steel producing country in the world. This sector represents around Rs. 90,000 crore of capital and directly provides employment to over five lakh people.

Displacement and Rehabilitation

It is true that the establishment of mega industrial projects was essential for development and to bring the economy of the country as developed economy. But establishment of industrial projects in tribal areas have brought the problem of displacement and uprooting of tribal families. Impact of industrialisation on tribes of Chotanagpur was studied by the following scholars:

1. The H.E.C. HATIA, Ranchi by Vidyarthi in 1970
2. The Patratu by Singh (1966)
3. The Bokaro by Sarkar (1970)
4. The Jamshedpur by Oraon 1958, UNESCO, Mishra 1959 and Sharma 1960.

The above studies done by anthropologists and scholars reveal the problem of uprooting and displacement among tribals.

HEC HATIA (Ranchi)

The study of Hatia by Vidyarthi reveals that the pre-industrial setting of present Hatia industrial zone was characterised by an all-round homogeneity in terms of landscape, population, economy and general style of life. The entire core of industrial Zone covering a total area of about 9200 acres was characterised by agricultural undulating land (Don) and upland (Tanr). Ethnically, it was populated by two major tribes—Ora on and Munda. Of total number of families which were uprooted in the villages, 77.5 per cent were tribals, 14.7 per cent Hindus and only 7.8 per cent Muslims. All uprooted villages were based on single crop cultivation with paddy as the chief crop. Due to insufficiency of agricultural production, people were forced to look for other means of supplementing their income. A large number of labourers from these villages used to go to Ranchi to work as casual labourers in off agricultural season.

The people of the uprooted villages were tradition bound. They had high regard for their supernatural dieties, *Bonga*, sarana and their dead ancestors. Their religious beliefs and practices infused a sense of homogeneity, solidarity and a strength in the mind of the people. The rhythm of religious life was tuned with their agricultural operations. The tribal villages were integrated with the religions in terms of village exogamy through inter village visits on occasion of festivals, *jatra*, and exchange of agricultural labourers and participation of villagers of the region in weekly market. These extension of relationships with other villagers were confined to a radius of 8 to 16 kms. which used to constitute micro region of such social, economic and political interactions.

Thus, the pre-industrial setting of present HEC and affluent industrial society was mainly characterised by natural landscape, folk and agricultural economy, social solidarity, group cooperation and copartnership. The tribal tradition was conducive to religious orientation. The society functioned on the principle of reciprocity and mutual obligations.

The first knowledge about industrial installations reached there when the initial survey of site selection was undertaken. The idea to install a large scale factory which involved uprooting of traditional style of life was not acceptable even to those families who were supposed to be relatively advanced. The survey records several movements and group activities by villagers to oppose the acquisition of villages, agricultural lands and installation of factories. However, in spite of resistant steps, the local people had to submit to the determination of the Government for installation of factory in the area in the name of general good of the nation.

These agricultural people when uprooted had to face manifold problems. The first was to find an alternative occupation to earn their livelihood. The second was related to the fixation of rate of compensation and then its realisation from the Government and above all proper utilisation of compensation money. Undoubtedly the affected and uprooted families were given reasonable compensation, but they had to face numerous social problems, cultural crises, economic disorganisation, and social disintegration. The problem of their disintegration was not tackled with adequate foresight and planning. Although economic rehabilitation was taken into account, but attention on cultural rehabilitation, exploitation oppression and deprivation was not taken into consideration

While the local traditional life was disintegrated during the formulatory period of industrialisation, the constructive phase of the Hatia industrial complex witnessed a large influx of industrial unskilled and skilled workers from different parts of India and also from Nepal. Those who came through recruitment board in HEC were accommodated in temporary quarters meant for them. But the large number of migrants who came to work for private contractors were sheltered either in neighbouring villages or in the colonies of temporarily built huts in open lands. These slum dwelling labourers distributed in village area of Hatia. Dhurwa and slum colonies constituted a total population of 32,550 persons. Large number of industrial labourers had come from industrial centres of West Bengal (Durgapur), Orissa (Rour- kela) Madhya Pradesh (Bhilai), Uttar Pradesh and Kerala. Such a vast mass of industrial workers migrating from one place to another in quest of suitable jobs posed a problem for planners and those interested in welfare. They had caused a new type of industrial nomadism. Such a group of migrants had also created a problem for employment of local people because former were preferred owing to their experience for similar work elsewhere. About two-third of slum dwellers were threatened with the sense of insecurity and suffer from wage cuts and irregular payment.

During the formulatory and constructive phases of industrialisation, the Hatia village, nearest to the factory site became the most important business centre for fulfilling the requirements of migrants and slum labourers. Along with the folk and rural traits of the village Hatia, numerous urban industrial characteristics got mixed and a new way of life emerged. The congestion in the village increased due to influx of migrants. Numerous hotels and variety of shops came into existence to cater to the needs of floating labourers. The contractors and truckowners also made arrangements for temporary accommodation in or near the village. On the whole all these commercial and industrial development in Hatia greatly disturbed the homogenous and rural life style of the village. Ideas and ideals, language and population, food habit and dress pattern, social and religious outlook and entire way of life were characterised by heterogeneity and industrial outlook as people from different places with different cultural background came to live in semirural village and around.

Most of displaced families invested their compensation money on purchasing agricultural land elsewhere. Most of these villagers were settled in new areas. After rehabilitation, they had became aware of modern fashion, radios, transistors and material goods. Noticeable changes in the dress of young men had occurred. Only few families were unable to face challenge in new settlement and preferred to settle in new surroundings containing their post way of life.

With the opening quarries, the traditional occupations of these villagers had also been affected. Village crafts were on decline. Landless labourers and small farmers had taken up quarry works. Income and expenditure of these families have considerably gone up.

Among the employed of the village 712 were Hindus and 348 tribals. The tribals had joined in large numbers as manual labourers. More men of tribal community were learning towards work in factory in preference to agricultural occupations. The commercial activities had gave up. Major shift in occupational structure was found from agriculture, agricultural labour, basket making, pottery, parched rice setting and weaving of 993 persons, 140 were now engaged in the same occupation. The main cause of shift was technological development and loss of agricultural land in those villages. The villages for all practical purposes had become a part of industrial complex now. They were now characterised by heterogenous life. They were good example of blending rural and industrial style of life.

The study of eight other villages located at a distance of two to twenty miles from the factory also showed interesting process of change caused by industrialisation. Though their traditional agricultural economy continued to be intact, they had accepted industrial work as their main subsidiary occupation, particularly the young men. The main cause of adopting industrial occupation was lower amount of per capita land occupied with low yield in agriculture. The distance was also a factor. Dependence on agricultural activities was now less among adjacent villages than in far off villages. Earning of family members had considerably increased in adjacent villages than in far off villages. However, villagers were of view that installation of factory was more beneficial to the outsiders.

These industrially affected villages showed the evidence of combining agricultural works with the industrial jobs in formulatory and constructive period when a vast gang of unskilled labourers were required for the construction of industrial units. But later these villagers were not absorbed in industrial work for want of skills. This frustrated those who were attracted towards industrial works to raise their standard of living. It was unfortunate that neither Government nor corporation administration could provide training to equip with skill and technical know how required in an emerging industrial complex.

Patratu Industrial Complex

The establishment of Patratu Industrial Complex has also brought the problem of displacement, rehabilitation of up rooted families and trends of emerging industrial society. Now let us discuss about them one by one.

Displacement: In case of Patratu Industrial Complex, 4450 acres of land was acquired which had affected 19 villages with 202 families. Out of 19 villages, 13 villages were affected totally or partially due to acquision of land for the dam. The remaining six villages were affected for a power plant and accessories. As Patratu was meant for a power plant having a dam to supply water and a small township to house the personnels employed there. It had not posed many problems as was noted in case of steel factories.

Rehabilitation: All 202 families were rehabilitated at Patratu in 100.7 acres of land acquired by the Government. The rate of compensation given to the affected families ranged between Rs. 166 to Rs. 2666 per acre according to different types of farm land. But the rate at which the land was being sold in the region was as high as Rs. 8000 to Rs. 12,000 per acre. The major difference in the rate of compensation received by the uprooted families and the prevalent rate in the area developed a sense of inhospitableness among the villagers towards the factory and the Government. The uprooted families had lost their main style of life, agriculture, and they always used to wait for industrial work. The neighbouring agriculturists had started using the land for producing cash crops vegetables and other items required by the emerging industrial society at Patratu.

Change: In the initial phase Patratu and other villages were isolated and barterminded. Now they had become money minded. Although money was known to them in the past, but it was never the monopolised source of exchange. But now money was everything. It had become the medium of exchange in the market for purchasing cloth, foods and luxury goods. Now it was also determining the status of an individual.

Under the process of industrialisation, the people have shown a considerable degree of change in their occupational structure and other spheres of their economic activities. There was decline in agricultural occupation. Villagers were attracted towards employment opportunities in the industry as semiskilled and unskilled labourers.

The process of industrialisation had not affected the traditional core of social structure in neighbouring villages. Some of villagers had to leave their home temporarily, their sense of oneness in family organisation had not been much affected. Workers from neighbouring villages like Patratu, Basda, and Labga used to return their home daily in the evening from the work site. But workers coming from far off villages like Hariharpur and Rarha had to live near the factory site. They stayed there for work for a week and returned their villages at the end of the week. 73 per cent workers of the Hariharpur used to pay a weekly visit to their home and 27 per cent used to pay a monthly visit to their house. In case of Rarha, 30 per cent workers used to pay a weekly visit, 25 per cent workers use to visit fortnightly, 40 per cent workers used to visit monthly and 5 percent workers used to visit yearly on certain specific occasion. At the factory sites workers of different villages used to stay in one camp. The village ties of friendship and kinship were maintained. The heads of the families of workers were respected. Workers used to pay their wages earned in the hands of heads of the family.

Social customs, religions beliefs and practices reflected a minimum transformation. The religious institutions of Hindus,

Muslims, and tribals coexisted. The annual festivals and their celebrations at Patratu were joined by all castes and tribes in integrated manners. The influence of traditional leaders like Pahan and Mahato was diminishing and a band of educated and politically conscious leaders was fast emerging. The villagers working in the factories had been influenced by labour leaders of unions working for the safe guard of rights of 8000 tribal and non-tribal workers. The law and order problem had also emerged

Bokaro Steel Plant

The installation of Bokaro Steel Plant had also resulted in a total acquisition of 32,390 acres of land for various purposes like main plant, township, dam, railway yards etc. Such an acquisition had displaced 6066 families of tribals and non-tribals. Out of 6066 families, 1159 families mostly Hindus were shifted to rehabilitation sites from their respective villages.

Agriculture was the primary occupation of about 94 per cent of the total displaced persons. But after their displacement they were compelled to abandon their traditional occupation. Some of families had bought land for cultivation at a distance of 10 to 15 miles from two rehabilitation sites. In the absence of other means of livelihood, the displaced persons had to look for non-agricultural works. Out of total affected persons only 625 persons were employed in Bokaro steel project of these, majority were recruited as unskilled workers, labourers, peons etc.

The majority of displaced persons visited the various work sites in order to get job under the contractors. The contractors preferred to employ the experienced immigrant workers. Such an attitude was reflected in the payment of wages of workers. The income of displaced persons had considerably gone down. After exhausting the compensation money they had no alternative but to dispose of their cattle and brass utensils and move in search of new avenue for their existence.

The traditional village institutions like caste affiliation, cycle of festivals and rituals and *Jajmani* system were completely disintegrated. All round depression and despair appeared in the life of the uprooted families.

The neighbouring villages were favourably affected by the Bokaro steel plant. The higher caste people of Chas, Jodhadih, Chandankewari and others situated nearby were in the look for jobs in factory since such a job ensured a permanent source of income. Members of lower caste families came to factory site every morning to work as labourers under contractors. They usually did so in off agricultural season. The location of factory had given them incentive to go for education. Education was considered necessary for getting jobs in the factory. Some of the villagers of Chas have been encouraged to open different kinds of shops at Chas.

Jamshedpur Industrial Complex

The establishment of Jamshedpur Industrial Complex has also brought the problem of displacement, rehabilitation and change in villages displaced and situated nearby the industrial complex. In a study, it has been extablished that 18 villages with an area of 16, 810.24 acres of land were occupied by TISCO as early as 1907.18 villages were inhabited by 615 resident land-holders. Out of which 329 were tribals (Bhumij 249, Santhal 66 and Ho 14) and remaining 306 were of different castes like Brahmin, Kamara, Suri, Gaur, Tanti, Dhana, Napit Kumhar, Dhobi, Kurmi, Rajput, Teli, Bagal, Ghasi and Sadgope.

The size of these villages varied from 6 to 25 families. The entire area was covered with jungle with agricultural land around the village. The people were dependent on agriculture. Their economy was subsistence oriented. They were mainly isolated from the world outside because of their geographical location between two rivers- swarnarekha and kharkai.

Villages were predominantly inhabited by tribal communities mainly Bhumij, Santhal and Ho. The percentage of tribal families in the villages constituted 60 per cent to 90 per cent. The percentage of tribal families in Sakchi and its two hamlets Kashidih and Mahulbera was cent-per-cent.

Earlier reports were silent on the rehabilitation and welfare of uprooted families by TISCO industrial complex. Due to absence of rehabilitation programme, the uprooted families took shelter as refugee in the areas lying outside the acquired belt. They took shelter in Bastis where other tribal migrants also came to settle. Thus, now-a-days, we have 33 such Bastis. The uprooted families were given preference in employment as unskilled labourers. Owing to unfamiliarity with jobs as well as fear complex, very few could take advantage of the opportunity given to them in the constructive phase of the industries. On the basis of study of 5 per cent sample families (Total No. 421) of tribal chosen from 16 Bastis, certain factors about cultural and economic transformation can be ascertained. Among the tribal labourers 13 per cent were of original displaced villages, while 87 per cent had migrated from the other parts of tribal areas to take up jobs in the city. As regards the period of migration 9.03 per cent had come in 1915-1929, 69 per cent during 1930-1944 and 22 per cent during 1945-1960. In terms of occupational category, 80 per cent were in unskilled category, 12 per cent were in skilled works and 7 per cent in ministerial work and 1 per cent in other types of work.

Although these migrants stayed in villages, but 97 percent of them maintained contacts with their parent villages and 3 per cent had severed all contacts with the villages. Some families of tribals also lived in company's houses which had amenities like tap water, electricity, and sanitation facilities. When these city dwelling tribals visited their villages, they found their life somewhat uncomfortable. Their children did not want to go to villages. The attitude and world view of city dwellers had undergone considerable change. The impact of urbanisation on tribals living in cities, *Bastis* and villages was observed, but somewhat different.

Rapid industrialisation in Jamshedpur had brought complete disintegration of pre-industrial culture characterised by forest ecology and homogenous folk style of life. The local agricultural economy suffered a complete break down. Local villagers had either to abondon the place or to get assimilated into the industrial way of life.

Conclusion

From the above studies of impact of industrialisation on tribals of Chotanagpur by the anthropologists, the following conclusions can be drawn:

1. The process of urbanisation has given birth to several mining, industrial, transportational and power towns in hilly and forested areas of Chotanagpur.
2. Industrialisation had led to a tremendous growth of means of communication and transport, investment of foreign capital, technical know how and expansion in to the world market.
3. There is considerable migration from rural to urban industrial centres due to urbanisations.
4. Influx of migrants in industrial centres is more the result of push factors due to low level of living condition in villages and lack of physical security.
5. Over population in and around industrial centres has brought the problem of unemployment, congestion, slum, sex and age disparity, inadequacy of civil amenities and above all lack of permanent band of skilled industrial workers.
6. Majority of migrants labourers used to maintain contacts with their respective villages. Only few of high income group workers had severed all connections with their respective villages.
7. Some labourers used to come to perform work from their villages from a distance of 5 to 10 kms on foot or on bicycles. It resisted reorganisation of industrial society and functioned as link between rural and urban migration.
8. Several big groups of migratory labour fance had emerged in initial phase of urbanisation for diging and construction works. Such migratory labour force wandered from place to place in search of industrial works.

9. In initial phase of industrialisation tribal and scheduled caste women from neighbouring villages provided big lobour force. But later hazards of hardship, uncertainties and difficulties of adjustment coupled with lack of opportunities discouraged migration of women. Sex disparity in industrial centres were also caused by migration of male members leaving their wives in their nativeplaces.
10. Sex disparity in urban centres brought the problem of social disorganisation. It created stress and strain in personal life as well as, sometimes, their personal demoralisation.
11. Monotonous, dull and drat life in factories, lack of recreational facilities through clubs, societies or cultural organisations contributed ultimately unhappiness, prostitution, alcoholism and other such vices which were rampant in industrial centres.
12. Industrial urbanisation had affected greatly the folk and tribal populations which till recently led a homogenous, distinctive and folk style of life. Their traditional way of life was disintegrated and they had to struggle hard to adjust themselves with the changed environment.
13. The large scale industrialisation brought the problem of displacement and uprooted families had lost their traditional economy, land house and traditional way of life.
14. Compensation was not given property. The lands were acquired at very cheap rate in comparison to existing rates.
15. Little compensation money was not fixed in banks. So, the compensation amount was spent on food, drink and fashion.
16. In absence of education and skill, members of uprooted families could not get employment in the industries established on lands acquired by them.
17. A number of uprooted families took shelter in slum areas. They were reduced to the states of labourers from the lord of land.
18. Uprooted family members could not compete even with

other migrant labourers who were more skilled and expert in industrial works.

19. Unemployment has brought frustrations in the family members of uprooted families. Instances of individual and family disorganisations have also been observed.
20. Contractors, moneylenders, traders, shopkeepers etc. have come in tribal *Bastis* and slum areas. Tribal families are indebted, exploited and oppressed.
21. Tribal women have been able to get work as Reja, Dai, Naukarani. Incidence of physical exploitation and prostitution has also been reported.
22. Improper rehabilitation of uprooted families has made a number of them instinct. Some of them have become slum dwellers. Some are leading their life worst than animals.
23. In the rehabilitation programme, cultural rehabilitation was not taken into account. Land by land or land by money compensation programme became failure due to lack of cultural rehabilitation programme for the uprooted families.

15

Role of Anthropology and Anthropologists in Development

Role of Anthropology in Development

Anthropology is science of man or study of man. This science makes study of man as a social, cultural, psychological, physical, linguistic, prehistoric and developmental being. There is no doubt, man is a developmental being. Man had made progress. He has passed through the successive stages of development like savagery, barbarism and civilisation. He is a progressive being and has made progress in different areas through inventions and discoveries. He is actively engaged in making scientific inventions. Development is for man, by man and of man. Thus, development is a significant activity of man. Anthropology as study of man deals with development activities of man too. Now we have developmental anthropology as a separate branch. Therefore, role of anthropology cannot be ignored in development. Role of anthropology in development can be discussed under the following headings:

1. Significant Role in Problem-finding

Anthropology being field science and applied value plays significant role in finding out the problem of development. As a field science, it can suggest about the target areas, target groups, beneficiaries and the general goal of development.

2. Significant Role in Problem Solving

The role of anthropology is not only significant for finding out the problem of development, but it also plays significant role in solving the problem of development. Being field science it can provide solution to the problem related with development. It can play decisive role by making the Government aware of the feeling of local people who are to be made developed.

3. *Can Highlight Cultural factors Supporting Development*

Anthropology being study of culture of mankind can highlight the cultural factors supporting the development. In culture bound society, development programmes cannot be successful without cultural acceptance. Cultural factors play significant role in supporting development. Anthropology can bring out these cultural values before the planners.

4. *Can Highlight Barriers of Development*

Being field science anthropology can highlight the barriers of development. It can place before the development planners a number of taboos, restrictions, belief, and practices prevalent in society which act as barriers of development. Knowing the facts, development barriers can be ended.

5. *Helpful to Policy Makers*

Anthropological knowledge about the culture and society of the people can be helpful to policy framers, plan formulators and others related to development strategy.

6. *Helpful to Development Administrators*

Anthropological knowledge about area and people can be of great help to the development administrators working with the people in field.

7. *Helpful to Development Programme Evaluation*

Anthropological knowledge is also helful to the persons or institutions associated with evaluation of development programmes.

8. Helpful to Programme Implementors

Anthropological knowledge, methods, techniques and approaches are also helpful to Governmental, and non-Governmental agencies associated with implementation of development plans.

9. Helpful to Communication Agencies of Development

Anthropological knowledge about cultural symbols, sound symbols and other kind of verbal and non-verbal behaviours extends help to communication agencies of development. The media agencies send messages about development to the people. While sending messages, they are symbols of the local people which make the message understandable to local people.

On the basis of above discussion, we can say with certainty that Anthropology can play significant role in development. Its significant role in development cannot be ignored.

Role of Anthropologists in Development

Not only anthropology but anthropologists, too, can play significant role in development. Anthropologists possess knowledge of Applied, Action and Developmental Anthropology. Therefore, they can contribute alot in the development process and planning. The role of anthropologists in development can be discussed under the following headings:

1. Role in Exploring the Problem of Deprived Communities

Anthropologists on the basis of their studies have been able to explore the problems of deprived communities such Scheduled Castes, Scheduled Tribes, women, children and aged. The problems explored by the anthropologists have been valuable for policy makers and planners.

2. Role in Highlighting Solutions to Development Problems

Anthropologists not only explore the problems faced by deprived communities, they also highlight the solutions for solving the problems of deprived communities. Anthropologists on the basis of their studies make recommendations to the Government for solving the problem.

3. Role in Development Strategy

Anthropologists have played significant role in formulating development strategy for the tribals. Anthropologists as chairman of review committee of different plans, or as members have been able to contribute significantly in development strategy. Verrier Elwin was chairman of committee on special multipurpose Tribal Blocks (1959-60). L.P. Vidyarthi was chairman of study group on Tribal Labour (1966-69). S.C. Dube was chairman of expert committee on Tribal sub-plan (TSP) set up by Ministry of Education and Social Welfare in 1972. Thus, anthropologists have played significant role in development strategy of Scheduled Tribes.

4. Role in Tribal welfare

Anthropologists have been associated with the Ministry of Tribal Welfare and Welfare activities of the tribals. They have formulated, completed and submitted their research project reports to the Ministry of Welfare for needful action. Their reports have been proved fruitful for development planners and policy makers.

Anthropologists have occupied chair in Tribal Research and Training Institutes. These research Institutes are integral part of the Tribal Welfare Ministry. By working as director or deputy director of the institute, anthropologists have played significant role in tribal development.

5. Role as applied Anthropology

Anthropology is a field based science and applied science. Therefore, anthropologists have to play the role of applied scientists in problem solving. They have been able to curb the concept of racial superiority, racial conflicts, tribal unrest etc.

during British period. They have also applied anthropological knowledge for the development and welfare of tribals.

6. Role as Action Anthropologist

Anthropologists have also played the role of an action anthropologists in making some tribal communities developed under their own supervision and work Prof. L.P.Vidyarthi had played significant role as an action anthropologist for making the Birhor Tribe developed. He had formulated, implemented and supervised the development programme of the Birhor, a primetive tribe of Jharkhand. Prof. P.K. Bhomick had formulated, implemented and supervised the development programme of Lodha, a tribe of West Bengal. These action anthropologists had played significant role in development of the tribe concerned.

7. Role as Trainer to Development Administrators

Anthropologists as academicians have played significant role in organising training programme to development administrators.

8. Role as Evaluator of Development Programme

On the basis of making an evaluative study of development programmes, anthropologists have able to recommend for follow up action or to stop the development programme or to launch an another development programme.

9. Role as Teacher of Development Anthropology

Anthropologists also play significant role as teacher of development anthropology.

Bibliography

Aiyapan, A. 1965: *Social Revolution in a Kerala Village*, Bombay: Asia Publishing House.

Ambastha, N.K. 1966: *A Critical Study of Tribal Education*. Delhi: S. Chand and Company.

Atal, Yogesh 1961: "The Cult of Bheru in Mewar Village" Ranchi: *Journal of Social Research*, Vol. 141.

Bailey, F.G. 1958: *Caste and Economic Frontiers*. Bombay: Oxford University Press.

Bailey, F.G. 1960: Tribe, Caste and Nation. Bombay: Oxford University Press.

Bailey, F.G. 1963: *Political and Social Change*: Orissa. Bombay: Oxford University Press.

Balaratnam, L.K. 1958: *Society in India*. Madras: Science Association.

Balaratnam, L.K. 1953: *Anthropology on March*. Madras: Book Centre.

Beals, A.R. 1962: *Gopalpur: A South Indian Village*. Bombay: Oxford University Press.

Beals, A.R. 1955: "Change in Leadership of a Mysore Village" in *Village India's*, Edited by M.N. Srinivas 1955.

Beals, A.R. 1955: "Interplay among Factors of Social Change in a Mysore Village". In *Village India* edited by Mckim Marriot. Chicago, 1955.

Beals, A.R. 1959: Leadership in a Mysore Village. In *Leadership and Political Institutions* edited by Park and Tinker, Oxford 1959.

Berreman, G.D. 1963: *Hindus of Himalayas*. Berkelay: California University Press.

Berreman, G.D. 1964: "Brahmanism and Shamanism in Rahar Village. In *Religion of South Asia* edited by Harper, University of Washington.

Berreman, G.D. 1969: "Urgent Anthropology in India" In *Urgent Researches in Social Anthropology*, Shimla, Indian Institute of Advanced Study.

Bhamick, P.K. 1963: *The Lodhas of West Bengal Kolkata*: Punthi Pustak.

Bose, N.K. 1929: *Cultural Anthropology*. Bombay: Asia Publishing House.

Bose, N.K. 1932: *Canons of Orissan. Architecture*. Kolkata, Ramanand Chaterjee.

Bose, N.K. 1949: *Structure of Hindu Society*. Kolkata: Vishwabharati Granthalaya.

Bose, N.K. 1967: *Culture and Society in India*. Bombay: Asia Publishing House.

Bose, N.K. 1969: *Problems of National Integration*. Kolkata: Allied Publisher.

Bose, N.K. 1970: *Tribal Life in India*.

Das, A.K. 1968: *Trends of Occupational Patterns through Generations in Rural Areas of West Benga*l. Kolkata: Cultural Research Institute.

Das, A.K. 1969: *The Totas*. Kolkata: Cultural Research Institute.

Das, A.K. and S.K. Banerjee 1962 (A): *Impact of Industrialisation on Life of Tribals* of West Bengal. Kolkata: Cultural Research Institute.

Das, A.K. and S.N. Banerjee. 1964: Impact of Tree Industries on *Tribals of West Bengal*. Kolkata: Cultural Research Institute.

Das, M.K. and M.K. Raha, 1963: *The Oruon of Sundervan*, Kolkata: Cultural Research Institute.

Das, M.K., M.K. Raha and B.K. Choudhury 1966: *The Mal Paharia of West Bengal*, Kolkata: Cultural Research Institute.

Das, M.K., M.K. Raha and B.K. Choudhury 1966: *Hand Book on Scheduled Castes and Scheduled Tribes of West Bengal*, Kolkata: Cultural Research Institute.

Das, K.R.S. 1966 " The Mina Tribe of Mewar" Kolkata: *Journal of Royal Asiatic Society of Bengal*. Vol. 55.

Das, K.R.S. 1967: *Study on Tribal People and Tribal Areas*. Indore: Government Regional Press.

Dube, S.C. 1993: *Understanding Change*; New Delhi: Vikas Publishing House.

Dube, S.C. 1990: *Traditions and Development*. New Delhi: Vikas Publishing House.

Elwin, Verrier 1939: *The Baiga*. London: John Murray.

Elwin, Verrier 1942: *The Agaria*. Bombay: Oxford University Press.

Elwin, Verrier 1943: *Maria: Murder and Suicide*. Bombay: Oxford University Press.

Elwin, Verrier 1947 (A): *The Maria and their Gothul*. Geoffery Cambridge: Oxford University Press.

Elwin, Verrier 1947 (B): *Myths of Middle India*. Geoferry Cambridge: Oxford University Press.

Elwin, Verrier 1951: *The Tribal Art of Middle India*. Mumbai: Oxford University Press.

Elwin, Verrier 1952: *The Loss of the Nerves: A Comparative Study of Contact of People in Aboriginal Area of Bastar State*. Central Province of India.

Elwin, Verrier 1954: *The Tribal Myths of Orissa*: Bombay: Oxford University Press.

Elwin, Verrier 1955: *The Religion of an Indian Tribe*. Geoferry Cambridge: Oxford University Press.

Elwin, Verrier 1958: *Myths of North East Frontiers of India*. NEFA: Shillong.

Elwin, Verrier 1959: *India's North-East Frontiers Agency in 19th Century*. Calcutta: Oxford University Press.

Elwin, Verrier 1960: *Philosophy of NEFA*. NEFA: Shillong.

Elwin, Verrier 1960: *Nagaland*. Shillong: Research Development Advertisers Secretariate.

Elwin, Verrier 1962: *Arts of NEFA*: Shillong: NEFA.

Ghurye, G.S. 1932: *Caste and Races in India*. London: Kegan Paul.

Ghurye, G.S. 1938: *Sex Habit of Middle Class People*. Pune: Gokhale Institute of Political and Economic Growth.

Ghurye, G.S. 1943: *The Aborigines so called and Their Future*. Pune: Gokhale Institute of Political and Economic Growth.

Ghurye, G.S. 1945: *Culture and Society*. Bombay: Popular Book Depot.

Ghurye, G.S. 1957: *The Mahadeva Kolies*. Bombay: Popular Book Depot.

Ghurye, G.S. 1959: *The Scheduled Tribes*. Bombay: Popular Prakashan.

Ghurye, G.S. 1960: *After a Century and a Quarter*. Bombay: Popular Prakashan.

Ghurye, G.S. 1961: *Caste, Class and Occupation*. Bombay; Popular Prakashan.

Ghurye, G.S. 1962: *Family and Kin in Indo-European Family*. Bombay: Popular Prakashan.

Ghurye, G.S. 1964: *Indian Sadhu*. Bombay: Popular Prakashan.

Ghurye, G.S. 1968 (A): *Religious Consciousness*. Bombay: Popular Prakashan.

Ghurye, G.S. 1968 (B): *Social Tension in India*. Bombay: Popular Prakashan.

Ghurye, G.S. 1973: *I and Other Explorations*. Bombay: Popular Prakashan.

Ghurye, G.S. 1974: *Wither India*. Bombay: Popular Prakashan

Ghurye, G.S. 1978: *India Recreates Democracy*. Bombay: Popular Prakashan.

Ghurye, G.S. 1979: *Vedic India*. Bombay: Popular Prakashan.

Ghurye, G.S. 1980: *The Burning Caldrom of North-East*. Bombay. Popular Prakashan.

Gores, G. 1938: *Himalayan Village: An Account of Lepchas of Sikkim*. London: Michael Joseph.

Grigson, W.G. 1939: *The Maria Gonds of Bastar*. London: Oxford University Press.

Guha, B.S. 1944: *Racial Elements in Indian Population*. London: Oxford University Press.

Guha, B.S. 1951: *The Tribes of India*. Two Volumes. New Delhi: Bhartiya Adim Sewak Sangh.

Guha, B.S. 1955: *Indian Aborigines and Who are They, Adivasi*. Delhi: Publication Division.

Haimendorf, Furer C. Von. 1939: *The Naked Naga*. London: McMillan and Company.

Haimendorf, Furer C. Von. 1941: "Seasonal Nomadism and Economics of Chenchus of Hyderabad". *Journal of Royal Asiatic Society of Bengal*, Volume. 7.

Haimendorf, Furer C. Von. 1943: *Chenchus: The Jungle Folk of Deccan*. London: McMillan and Company.

Haimendorf, Furer C. Von. 1945 (A): *The Tribal Population of Hyderabad*. Hyderabad: Revenue Department.

Haimendorf, Furer C. Von. 1945 (B): *The Redies of Beson Hills: A study in Acultuation*, Two Volumes. London: McMillan and Company.

Haimendorf, Furer C. Von. 1946 (A): "The Agriculture and Land Tenure Among the ATaPani": *Man in India*. Vol.26.

Haimendorf, Furer C. Von. 1946 (B): "A Notes on Tribal Justice Among the Apatanis".

Haimendorf, Furer C. Von. 1947: *Explorations in Eastern Himalayas*. Shillong: Assam Government Press.

Haimendorf, Furer C. Von. 1948: *The Raj Gond of Alidabad: A Peasant Culture of Deccan*. London: McMillan and Company.

Haimendorf, Furer C. Von. 1950-51: "The Pradhans: Bardes of Raj Gond". *Eastern Anthropologist*, Vol.4.

Haimendorf, Furer C. Von. 1955: *The Himalayan Barbery*. London: John Murray.

Haimendorf, Furer C. Von. 1962: *The Apatani and Their Neighbours: A Primetive Civilisation of Eastern Hemalayas*. New York. Kegan Paul.

Haimendorf, Furer C. Von. 1964: *The Sherpas of Nepal*. London: Kegan Paul.

Haimendorf, Furer C. Von. 1967: *Morals and Merits*. London: Kegan Paul.

Haimendorf, Furer C. Von. 1969: *The Moniac Nagas*. London: Kegan Paul.

Haimendorf, Furer C. Von. 1979: *The Gonds of Andhra Pradesh*. London: Kegan Paul.

Halder, A.K. and Gaya Panday 2002: *Aritsan Tribes of Bihar and Bengal*. Kolkata: North Bengal Press.

Hutton, J.S. 1921 (A): The Angami Naga. London: McMillan and Company.

Hutton, J.S. 1921 (B): *Sema Naga*. London: McMillan and Company.

Iswaran, K. 1966 (A): *Traditions and Economy in Village India*. London: Routledge and Kegan Paul.

Iswaran, K. 1966 (B): *Shivapura: A South Indian Village*. London: Routledge and Kegan Paul.

Iswaran, K. 1968: *Caste in South Indian Village*. London: Routledge and Kegan Paul.

Kapadia, K.M. 1968: *Marriage and Family in India*. Calcutta: Oxford University Press.

Karve, Iravati 1957: 'Social Education of Scheduled Tribes', *Vanya Jati*. Vol. 3, No. 4.

Karve, Iravati 1958: 'Bhills of *Western Khandesh*'. Bombay: *Journal of Anthropological Society*.

Karve, Iravati 1969: 'Tribals of Maharashtra' In Urgen Researches in Social Anthropology: Shimla: Indian Institute of Advanced Study.

Kolenda, P.M. 1959: Towards a Model of Hindu Jujmani System. *Hindu Organisations*. Vol. 22 No-1.

Lewis, Osker 1954: *Group Dynamics in a North Indian Village*. New Delhi: Planning Commission.

Lewis, Osker 1958: *Life in a North Indian Village*. Durban: University of Illinois Press.

Majumdar, D.N. 1937: *A tribe in Transition: A Study in Cultural Pattern*. London: Long Mans Green and Company.

Majumdar, D.N. 1944 (A): *A Fortunes of Primetive Tribes*. Lucknow Universal Publishers.

Majumdar, D.N. 1944 (B): *Races and Culture of India*. Allahabad: Kitabistan.

Majumdar, D.N. 1950: *The Affair of a Tribe: A Study in Tribal Dynamics*. Lucknow: Universal Publisher.

Majumdar, D.N. 1958: *Caste and Communication in an Indian Village*. Bombay: Asia Publishing House.

Majumdar, D.N. 1960 (A): *Chorka Ek Gaon*. Bombay: Asia Publishing House.

Majumdar, D.N. 1960 (B): *Social Contours of an Industrial City*. Bombay: Asia Publishing House.

Majumdar, D.N. 1962: *Himalayan Polyandry*. Bombay: Asia Publishing House.

Mathur, K.S. 1964: *Caste and Ritual in Malwa Village*. Bombay: Asia Publishing House.

Mayer, A.C. 1957. *Land and Society in Malawar*. Bombay: Asia Publishing House.

Mandelbaum, David A. 1938: 'Polyandry in Kota Society'. *American Anthropologist*, New Series Vol. 40.

Mandelbaum, David A. 1941: 'Culture Change Among Nilgiri Tribes'. *American Anthropologist*, New series, Vol. 4.

Mandelbaum, David A. 1955 (A): 'Technology Credit and Culture in a Nilgiri Village' in *India's Village* Edited by M.N. Srinivas.

Mandelbaum, David A. 1955 (B): 'The World and World View of Kota'. In *Village India* edited by Mckim Marriot.

Marriot, Mckim (ed.) 1955: *Village India*. Chicago: *American Anthropologist*, Vol. 57.

Marriot, Mckim 1960: *Caste Ranking and Community Structure in Five Regions of India*. Puna: Deccan College.

Mishra, N. 1971: *Culture Persistence and Change*. Delhi: Concept Publishing Company:

Mishra, N. 2003. Dalit Exploitation in Bihar. Delhi: D.K. Publishers.

Naik, T.B. 1964: *Economic Organisation and Market*. Chhindawara: Tribal Research Institute.

Oruon, Prakash. 2002 : *Land and People of Jharkhand*. Ranchi: Jharkhand Tribal Welfare Research Institute.

Pandey, G. 2006: *Manavshastriya Anusandhan Vidhi Ewam Taknik*. New Delhi: Concept Publishing Company.

Pandey, G. 2006: *Jatil Samaj Ka Manav Shastra*. New Delhi: Concept Publishing Company.

Pandey, G. 2006 : *Bhartiy Manav Shastra*. New Delhi: Concept Publishing Company.

Pandey, G. 2007. *Manavshastriya Sidhanta: Shastriya Ewam Adhunik*. New Delhi: Concept Publishing Company.

Pandey, G. 2007: *Bhatiya Janajatiya Sanskriti*. New Delhi. Concept Publishing Company.

Presad, N. 1964: *Land and People of Tribal Bihar*. Ranchi: Bihar Tribal Research Institute.

Radcliffe-Brown, A.R. 1922: *The Andaman Islanders*. Glencoe Illinois: Oxford University Press.

Redfield, Robert. 1955: *The Little Community: View Point for the Understanding Human Whole*. New York: Upasal and Stockwholem. Department of Education.

Report 1998: Ministry of Human Resource Development, Government of India, New Delhi.

Report 1999: Department of Education, Ministry of Human Resource Development. New Delhi: Government of India.

Report 1998: Ministry of Health and Family Welfare. New Delhi: Government of India.

Report 1999-2000: Department of Women and Child Development, Ministry of Human Resource.

Report 1999: New Delhi: Government of India. Planning Commission, Government of India, New Delhi.

Report 1998-99: Ministry of Social Justice and Empowerment

Report 2000-01: Ministry of Tribal Affairs. New Delhi: Government of India.

Rivers, WH.R. 1906: *The Toda*. London. Mcmillan and Company.

Roy, S.C. 1912: *Munda and Their Country*. Ranchi: City Bar Library.

Roy, S.C. 1914: *The Oraon of Chotanagpur*. Ranchi: City Bar Library.

Roy, S.C. 1925: *The Birhar: Little Known Jungle Tribe of Chotanagpur*. Ranchi: Man in India Office.

Roy, S.C. 1928: *Oraon Religion and Custom*. Ranchi: Man in India Office.

Roy, S.C. 1935: *The Hill Bhuiyas of Orissa*. Ranchi: Man in India Office.

Roy, S.C. 1937: The Kharia. Two Volumes. Ranchi: Man in India Office.

Sachidananda 1964: *Culture Change in Tribal Bihar*. Calcutta: Book Land Pvt. Ltd.

Sachidananda 1966: *Profile of Tribal Culture in Bihar*. Calcutta: Bookland Pvt. Ltd.

Sachidananda 1968: *Tribal Village in Bihar: A Study in Unity* and Extension. Delhi: Munshiram Manohar Lal.

Sachidananda 1976: *Tribal Voters in Bihar*. Calcutta: Bookland Pvt. Ltd.

Sachidananda 1977 (A): *Sarvodya and Development*. Calcutta: Bookland Pvt. Ltd.

Sachidananda 1977 (B): *Harijan Elite*. New Delhi: Concept Publishing Company.

Sachidananda 1980: *Changing Munda*. New Delhi: Concept Publishing Company.

Sachidananda 1984: *Aspects of Woman in India*. New Delhi:

Concept Publishing Company.

Sachidananda 1986: *Scheduled Castes*. New Delhi: Concept Publishing Company.

Sachidananda 1987 (A): *Industrialisation and Social Disorganisation*. New Delhi: Concept Publishing Company.

Sachidananda 1987 (B): *Swidden Cultivation in India*. New Delhi: Concept Publishing Company.

Sachidananda 1997: *Tribal Study Series*. Delhi: Serial Publications.

Sachidananda 2001: *Dalit Women on Road*. Delhi: Serial Publications.

Sahay, K.N. 1975: *The Hindu Shrines of Chotanagpur*. Shimla: Indian Institute of Advanced Study.

Sahay, K.N. 1976: *Under the Shadow of Cross*. Calcutta: Institute of Social Research and Applied Anthropology.

Sahay, K.N. 1977: *Christianity and Culture Change*. Ranchi: Basant Press.

Sahay, K.N. 1993: *Visual Anthropology*. Delhi.

Sahay, K.N. 1994: *Interactions of Cultural Traditions in India*. Delhi.

Sahay, K.N. 1998 (A): *Dynamics and Dimensions of Tribal Societies in India*. Delhi.

Sahay, K.N. 1998 (B): *Anthropological Study of Cartoons in India*. Delhi.

Sahay, K.N. 1998 (C): *Spiritual Anthropology*. Delhi.

Sahay, Sushma, Prasad 1988: *Tribal Women Labourers: Aspects of Economic and Physical Exploitation*. New Delhi: Gyan Publishing House.

Sahay, V.S. and P.K. Singh: *Indian Anthropology*. Allahabad: K.K. Publications.

Saxena, K.B. 1983: *Tribal Development in Bihar*. Ranchi: Tribal Welfare Research Institute.

Sharan, A.B. 1975: *Murder and Suicide Among Three Tribes of Chotanagpur*. New Delhi: National Publishing House.

Sharma, B.D. 1976: *Tribal Development: Conceptual Frame*. Bhopal: Adim Jati Research and Development Institute.

Sharma, B.D. 1986: *Adivasi Vikas: A Saidhantic Vivechan*. Bhopal: Madhya Pradesh Hindi Grantha Academy.

Shukla, R.K. 1998: *Forestry for Tribal Development.* New Delhi.
Singh, J.P., M.N. Vyas and R.S. Mann 1995: *Tribal Women and Development.*
Srinivas, M.N. 1952: *Religion and Society among Coorgs of South India.* Bombay.
Srinivas, M.N. 1955: *India's Village.* Bombay.
Srinivas, M.N. 1959: *Methods in Social Science.*
Srinivas, M.N. 1962 (A): *Caste in Modern India.* Bombay.
Srinivas, M.N. 1962 (B): *Caste in Modern India and Other Essays.* Bombay.
Srinivas, M.N. 1966: *Social Change in Modern India.* Bombay.
Srinivas, M.N. 1976 (A): *The Remembered Village.* California.
Srinivas, M.N. 1976 (B): *Nation Building in Independent India.* Oxford.
Srinivas, M.N. 1986: *Dominant Caste and Other Essays.* Delhi.
Srinivas, M.N. 1989: *The Cohesive Role of Sanskritisation,* Delhi.
Srinivas, M.N. 1992: *On Living in a Revolution.* Delhi.
Srinivas, M.N. 1996: *Village, Caste, Gender and Method.*
Srinivas, M.N. 1997: *Caste its Twentieth Century Avatar.* Delhi.
Thapar, Ramesh (ed.) 1977: Tribe, Caste and Religion in India. Delhi: MacMillan India Pvt. Ltd.
Upadhyay, V.S. and G. Pandey 1993: *History of Anthropological Thought.* New Delhi: Concept Publishing Company.
Upadhyay, V.S. and G. Pandey 2001: *Manavshastriya Vicharak Ewam Vichar Dharayen.* Delhi: Directorate of Hindi Medium Implementation. Delhi University.
Upadhyay, V.S. and G. Pandey 2002 (A): *Vikasatmak Manav Vigyan.* Bhopal: Madhya Pradesh Hindi Grantha Academy.
Upadhyay, V.S. and G. Pandey 2002 (B): *Janajatiya Vikas.* Bhopal: Madhya Pradesh Hindi Grantha Academy.
Upadhyay, V.S. and G. Pandey 2002 (C): *Tribal Development in India.* Ranchi: Crown Publications.
Upadhyay, V.S. and G. Pandey 2004: *Samajik Sanskritik Manav Shastra.* Ranchi: Crown Publications.
Verma, R.C. 1997: *Bhartiya Janajatiyan: Atit ke Jharokhe Se.* New Delhi: Publications Division, Government of India.
Verma, U.K. 1991: *Bihar Ka Janajatiya Jivan.* Varanasi: Kishore Vidya Niketan.

Vidyarthi, L.P. 1961: *Sacred Complex in Hindu Gaya.* Meerut: Kedar Nath Ram Nath.

Vidyarthi, L.P. 1963: *The Malar: A Study in Nature-Man-Spirit Complex of Hill Tribe.* Calcutta: Bookland Pvt. Ltd.

Vidyarthi, L.P. 1969: *Cultural Configuration of Ranchi.* Calcutta: Bookland Pvt. Ltd.

Vidyarthi, L.P. 1973: *Dynamics of Tribal Leadership in Bihar.* Allahabad: Kitab Mahal.

Vidyarthi, L.P. 1976: *Tribal Culture of India.* New Delhi: Concept Publishing Company.

Vidyarthi, L.P. 1977: *Rise of Anthropology in India* in Two Volumes. New Delhi: Concept Publishing Company.

Vidyarthi, L.P. 1980: *Tribal Development and Its Administration.* New Delhi: Concept Publishing Company.

Index